REPAIR
FINISHING
REFINISHING
BY ALLAN E. FITCHETT

Professional Techniques for Everyone

CONSTANTINE

TM

Albert Constantine & Son, Inc.
New York, NY 10461

IMPORTANT NOTICE

All practices and procedures mentioned and described in this book are effective methods of performing specific operations. All tools and materials mentioned will produce the most gratifying results when the manufacturers recommendations and instructions accompanying such tools and materials are followed. All safety precautions must be followed at all times. The latest available authoritative information is offered and contained in this book.

The publisher, Albert Constantine & Sons, and the author, Allan E. Fitchett, can not be held responsible for any personal injury or property damage.

Library of Congress Catalog Catalog Card No. 89-90757
ISBN 0-910432-00-7

DEDICATION

This book is dedicated as a memorial to my late wife, Jane. Her love, understanding, inspiration, encouragement and assistance during 43 years, enabled me to turn a 60 year old avocation into a vocation. Our three children, Maurine, George and Martha also provided an incentive for me to impart my experience, knowledge and abilities to others through lectures, demonstrations and writings. Without all this love, understanding and encouragement, this book would not exist.

TABLE OF CONTENTS

> Woods, Stripping Materials, Solvents, Adhesives, Abrasives, Tack Cloths, Coloring Materials, Stains, Filling Materials, Finishes, Rubbing & Polishing Compounds

> Washing and Cleaning, Stripping, Bleaching

> Joining Wood Surfaces, Types of Joints, Dowels and Doweling

> Replacing Broken Rungs or Spreaders, Tightening Loose Joints, Boring Out Old Dowels, Wood Selection, Clamping, Bending Wood

> Flattening Veneers, Broken or Missing Veneer, Repairing Blistered Veneer, Repairing Buckled Veneer, Removing Dents in Raw Wood, Filling Cracks and Holes

> Using Hand Scraper, Hand and Power Sanding

FOREWORD

The materials and techniques used for achieving a specific restoration can become very controversial especially when dealing with "Antiques." There are those who insist on using the same type tools, adhesive, stain, finish, and hardware that were used originally. That is their choice. I prefer to use modern type materials and tools which will produce the same visual results plus give you the satisfaction of knowing the piece will be usable and have greater longevity by incorporating the results of modern chemistry and technology.

The primary aim of this book is to acquaint those who are interested in repairing, finishing or refinishing a wooden surface, whether it be of old or new wood, with some of the latest tools, materials and techniques. Volumes of text material could be written about any one specific item or subject relating to working with wood. I have endeavored to condense the information contained in this book, relating only to the most often asked questions, the most often misunderstood information and the reasons for disastrous results that I have encountered in my associations with woodworkers over the years. The remedies and techniques offered will often have many applications and should not be construed to be confined to one specific item only.

My personal association with wood has often presented a challenge in construction, repair, finishing and refinishing of an assembly of pieces of wood. Using only a minimum amount of readily available tools and hardware, combined with quality adhesives and finishing materials, I hereby offer you the opportunity to fashion a most acceptable assembly of wood, one that will exhibit the beauty of the wood and be long lasting and one that you can be proud of. Some elements that can not be purchased but must be acquired are **common sense, ingenuity** and **patience.**

The techniques described in the following pages are the results of trial and error over many years and should not be construed to be the only way to achieve a certain result. They are the ways I have found to provide me with the most successful results with the least amount of equipment, materials and effort.

In addition to the text material relating to specific materials and techniques, it is strongly recommended that reference be made to the section, "The Materials We Work With." This will contain the **latest available information** pertaining to specific commercially available products and materials that you might consider for possible use in your endeavors.

1

Modern research and technology combined with aggressive advertising has offered the consumer many one-step, one-shot, quick, do-it-yourself repair and refinishing kits. Use them if they satisfy your requirements. Properly used, they will present an acceptable appearing finished surface.

The steps, procedures and materials described in this book are time tested. It will take you longer to achieve the final results, but there is no short cut or substitute for a quality, long lasting job, one that you can always be proud of.

By doing the work yourself, you will be constantly aware of exactly what went into the project and what you can expect from your efforts. You will have complete control over the entire situation at all times. When completed, take pride that you did it yourself. You have added character to an assembly of wood and since there are no two pieces of wood exactly alike, you have created an original, which can never be duplicated.

A book such as this would be impossible to compose without mentioning specific names of brands or manufacturers. Much help and advice was solicited and offered by those mentioned in the "Acknowledgements." The mere mention of such names or photographs of specific products is not intended as an endorsement, nor is the lack of names or photographs intended to discriminate against certain products or manufacturers.

The "Tips & Techniques" following each subject provide a very brief summary of the salient information relative to the immediate preceding subject.

Be aware that modern research and technology in the production and manufacturing of finishing materials is constantly changing, and at a rapid rate. The skillful use of these materials is not alone sufficient for those interested in the finishing process. The complete understanding of these materials and also the materials upon which you place them will be most rewarding. Be sure to read the instructions on the containers, which are often updated. Certain phrases which you will find repeated throughout the text, especially those referring to safety and testing, can not be overemphasized.

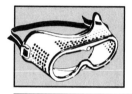

Safety
Make it a Habit

Most woodworkers are complacent individuals and are prone to state that potential hazards can't happen to them. Baloney! Personal experiences, and experiences of other woodworkers that have been related to me, prove that there are no short cuts for eliminating potential hazards and ignoring safety precautions. The very nature of the work permits woodworkers to be constantly exposed to potential hazards from the tools and materials they use in the creation of their Masterpieces.

Volumes of material have been written about precautions and proper use of both hand and power tools, each tool having its own particular hazard. It would be most advantageous to occasionally review the instruction booklet relative to the specific tool or machine you are operating. Pay particular attention to the safety instructions.

The basic material we use, Wood, is not generally considered to be a highly toxic material. However, exposure to the wood dust, from either sawing or sanding, can cause serious respiratory ailments, skin irritations or eye allergies. Be aware that each individual person has their own specific level of resistance to the potential exposure of the toxic and/or allergy producing materials. This level of resistance can also change as your age advances.

Studies and reports made during recent years have alerted the woodworker to the potential hazards from certain woods or wood products. A few of the wood species that are known that can possibly cause allergenic reaction to some humans are; Rosewood, Yew, Boxwood, Cashew, Satinwood, Teak, Ebony, Cocobola, Mahogany and Western Red Wood.

Though it may be that a specific tree, a milled board from that tree, a plane shaving or sawdust chip from the board can and will cause you some physical discomfort, it is most likely that the dust created during the cutting and sanding, which you can inhale into your respiratory tract, will produce the most severe problems to your nose, larynx, tonsils, tongue and lungs.

A few precautions that might be considered for the reduction of this wood dust hazard are: substituting a tool that produces less dust, such as using a plane instead of a belt sander; using a dust collector such as a "Shop Vac" or a vacuum cleaner with the nozzle of the hose placed near the source of the wood dust as it is being generated; or wearing a Filter mask that will at least separate the solid particles of wood out of the air you are breathing.

Filter Mask

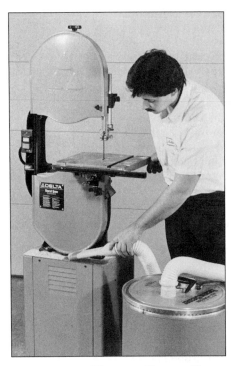

All-purpose Vacuum Cleaner

Faceshield

Aside from the wood and wood products themselves that woodworkers are exposed to, perhaps more potentially hazardous are the materials and vapors emitted from the adhesive materials that are used to bond the pieces of wood together and the materials we elect to finish the Masterpieces with, such as paint, paint and varnish removers, bleaches, stains, sealers, fillers, finishes and solvents. All these materials are chemical compounds and when exposed to the atmosphere, they readily emit vapors. Some are unseen and not very odoriferous while others give forth visible vapors and are odoriferous. These vapors are often flammable, toxic and more hazardous and serious than the wood dust itself.

The odor alone of a substance should not be used to evaluate the toxic effect of a specific material. Epoxies emit mild, odoriferous, very toxic vapors while highly odoriferous Acetone is very low on the toxicity scale. Just because a product is offered for sale to the average consumer, does not mean that it can not present a health hazard.

BE SURE TO READ THE WARNING LABELS ON ALL CONTAINERS

Also be aware that often the contents can be changed from time to time and the labeling information may have been updated.

The potential hazard from such toxic materials can be minimized by providing adequate ventilation in your work area; also the use of an organic type respirator is strongly suggested. It is true that wearing a respirator will not enhance your looks. It will restrict your field of vision and give you some difficulty in talking. Breathing will require more effort than normal. As you get accustomed to wearing a respirator, these problems soon become less noticeable. The benefits of using a respirator far outweigh the reasons for not wearing one. At the end of the day you will find you will be free from clogged nostrils, rasping coughs and gummy throats. The vapors will no longer give you that vague "hung-over" feeling. Gone will be the headaches and respiratory irritations that these materials and solvents could cause.

Respirators are available in many models, sizes and styles. It will be well worth your while to investigate the various types available and purchase one that will adapt to your requirements.

Dual Purpose Respirator

Heavy Duty Rubber Gloves

REMEMBER, THE LIFE YOU SAVE MAY BE YOUR OWN

In addition to the potential hazards to ourselves from the materials that we are working with, there is also the possibility of indirect hazards due to the accumulation of time and materials.

Spontaneous combustion can and does happen. Immediately after using any cloths with stains, finishing material or solvents, place the soiled cloth in a pail of water, preferably a metal pail with a metal cover. At the end of the day, take the pail with the soiled rags outside and away from any building or other combustible materials. Make proper and final disposition of all discarded materials as soon as possible.

Explosions can and do happen. When working in an area that is not well ventilated, the vapors from the materials you are working with can accumulate. If the vapors are flammable, a spark from the ignition of a furnace, a defective or improper switch, or a spark from an electric motor, could result in an explosion and the end results could be disastrous or even fatal.

When not being used, keep all containers properly covered, labeled and stored away from excessive changes in temperature, humidity or dampness, and potential sparks or flame.

Keep a first aid kit handy, along with the telephone number of your doctor and hospital.

Have a properly rated fire extinguisher handy along with the telephone number of your local fire department.

KINDS OF FIRE EXTINGUISHERS NEEDED IN A WOOD SHOP

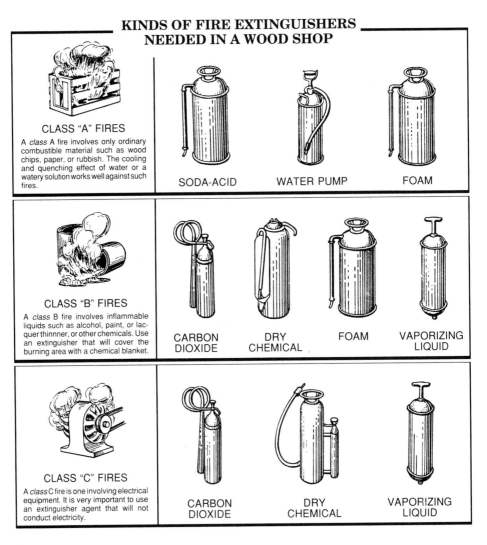

CLASS "A" FIRES

A *class* A fire involves only ordinary combustible material such as wood chips, paper, or rubbish. The cooling and quenching effect of water or a watery solution works well against such fires.

SODA-ACID   WATER PUMP   FOAM

CLASS "B" FIRES

A *class* B fire involves inflammable liquids such as alcohol, paint, or lacquer thinnner, or other chemicals. Use an extinguisher that will cover the burning area with a chemical blanket.

CARBON DIOXIDE   DRY CHEMICAL   FOAM   VAPORIZING LIQUID

CLASS "C" FIRES

A *class* C fire is one involving electrical equipment. It is very important to use an extinguisher agent that will not conduct electricity.

CARBON DIOXIDE   DRY CHEMICAL   VAPORIZING LIQUID

It is not the intent of this author to scare any woodworkers away from their craft or hobby. During my many years in woodworking, I have experienced or witnessed what can happen if safety is ignored. Modern technology and research has provided us with better tools, materials and information than we had in the past. We are now able to use tools and materials in a safer atmosphere, and produce longer lasting Masterpieces under many varied conditions.

The research and information that has been accumulated over the years has been documented, updated and finally submitted to the public for acceptance, and in many instances these results are being mandated by law. As this book goes to press, OSHA, EPA, etc. have issued regulations that will demand our immediate attention. Be alert for these changes.

The hazards mentioned above are only a few of the potential risks that the woodworker is constantly encountering. Diligent use of common sense and caution should allow the woodworkers to proudly continue to produce Masterpieces with safety to themselves and others who surround them. Most short cuts wind up with disastrous results.

Only recently, through public neglect of safety practices, though not related to wood, I lost my most precious possession, my wife of forty three years. Yes, I know what safety is and what neglect of it can do.

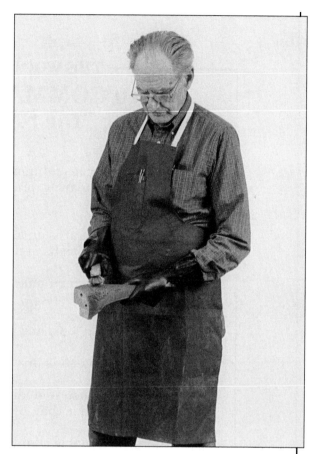

THE WOODWORKER'S
10 COMMANDMENTS
for SAFETY

1 Use protective equipment whenever necessary; eye protection, hearing protection, respirators, hardhats, etc.

2 Use proper guards on all power equipment plus proper operating positions and procedures.

3 Operate power equipment only after reading or receiving proper instructions.

4 Inspect hand tools and keep them in good condition.

5 Store all materials in a stable and secure manner.

6 Eliminate dust and dirt in the area by vacuum, if possible, rather than blowing or sweeping.

7 Store flammable liquids in suitable containers or cabinets.

8 Store combustible waste material in covered metal containers and dispose of it daily.

9 Have adequate ventilation particularly in finishing and drying areas.

10 Know where first aid can be found and where fire extinguishers are located.

Is It Worth Fixing ?

There is a great tendency today to acquire old or antique furniture that may have been in storage for many years. It may have been a piece subjected to daily or excessive use or which had been misused and/or abused over the years. Seldom is it hopeless to attempt a repair of a piece of old furniture. Don't be too quick to relegate the relic to the fireplace. By using a minimum of tools and materials, plus a portion of common sense, ingenuity and patience, you can restore this piece into a once again valuable, useable piece of furniture. In most cases it will be far superior in construction to the furniture made and purchased today and will have great longevity.

A Stickley Lamp

It may be a piece of furniture that was custom made for a family member by a family member and has acquired sentimental value over the ages. However, due to neglect of proper care and maintenance, the piece may no longer be solid or sturdy and is now in need of repairs and refinishing. You may have a piece of contemporary furniture that has been damaged. Whatever the case, you should decide how you want the piece restored and made useable once again.

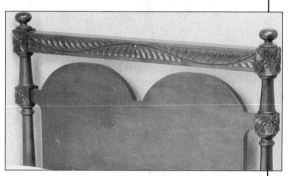

Carved Walnut Headboard

Do you want to send it out and have the work done by a professional? In this case, you should obtain estimates from at least three reputable professionals and have a written, signed contract of what is going to be done, how long it will take, what the cost will be and what guarantee will be provided for the future of the finished item.

Perhaps you want to attempt to try repairing and restoring it yourself. In this case you should have some knowledge or guidance in the proper procedures, tools and materials that are necessary and available to restore the piece to a condition that you can proudly be satisfied with. If you elect to attempt the restoration yourself, many books, courses and qualified consultants are available to assist you in your endeavors.

Pot Table

For the person who elects to tackle the job personally, you will have complete control over all the steps or procedures at all times and can evaluate the results as you progress. You may alter any of these steps, when necessary, as you continue through the process of repairing and finishing or refinishing your project. You will also have a choice to use whatever quality materials and tools you want and you will know exactly what is being done to the piece at all times. Finally, you will have the satisfaction when all is completed; you can take pride in the fact that you did it yourself.

Drop Leaf Table

Oak Side-by-Side

Mission Style Oak Rocker

The Materials We Work With

- WOODS
- STRIPPING MATERIALS
- SOLVENTS
- TYPES OF ADHESIVES and THEIR PROPERTIES
- ABRASIVES
- TACK CLOTHS
- COLORING MATERIALS
- STAINS
- FILLING MATERIALS
- FINISHES
- RUBBING and POLISHING COMPOUNDS

The following section contains a short description of some of the most common materials we work with. A brief background of these materials is presented along with some of the varieties of materials that you can consider using for your own specific application. Information and instructions about specific or special materials is provided when related, throughout the text. Carefully select the material that will work for you. What works in one instance may not work in another.

Above all, if you are not sure of the outcome of using a certain material, test it out on a scrap of the same type of wood you will be ultimately working on. Test your entire finishing schedule first. If you are not satisfied with the end result, you can always back up and make the necessary adjustments before advancing to your final project. Take care, be patient, have fun, and be proud of your results.

WOODS

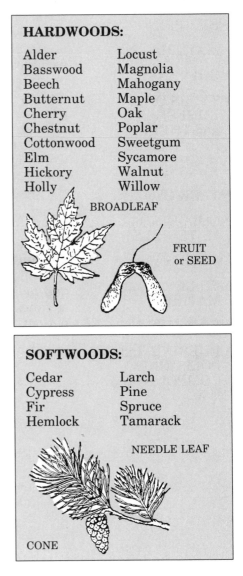

It is not necessary to be able to identify the 53,000-plus different kinds of wood; however it will be most helpful to be able to confer intelligently with your suppliers when ordering wood. Hence, the possession of a small wood identification kit, which you can always add to, will be most valuable. It will supply you with authentic, identified samples of the most common and exotic woods used in the furniture and wood industry and enable you to compare the identified samples with the woods you are working on and to then talk with confidence.

HARDWOODS and SOFTWOODS

Trees are divided into two broad, general classes, usually referred to as Hardwoods and Softwoods. Some softwoods, however, are actually harder than some of the hardwoods, and some hardwoods are softer than the softwoods. Botanically, the softwoods fall into a classi-fication of trees called Conifers that have their seeds exposed, usually in cones. Softwoods have needle-like or scalelike leaves that remain on the trees throughout the year. Hardwoods have true flowers and broad leaves, and their seeds are enclosed in a fruit. The hardwoods lose their leaves in the fall or during the winter.

HARDWOODS:

Alder	Locust
Basswood	Magnolia
Beech	Mahogany
Butternut	Maple
Cherry	Oak
Chestnut	Poplar
Cottonwood	Sweetgum
Elm	Sycamore
Hickory	Walnut
Holly	Willow

BROADLEAF

FRUIT or SEED

SOFTWOODS:

Cedar	Larch
Cypress	Pine
Fir	Spruce
Hemlock	Tamarack

NEEDLE LEAF

CONE

Each variety and type of wood is unique, having its own particular grain, color, texture, density, tensile strength, resistance to decay and its own working qualities.

Wood changes according to the part of the tree from which it was cut, how it was milled and how it was seasoned or dried. Some wood tends to twist or warp as it dries. Stresses in the wood or uneven moisture content can cause these distortions long after the wood is milled. Some wood has a tendency to develop splits or checks and some wood retains a lot of moisture. If there is too much moisture in the wood, it will be difficult to work with and will not glue up properly. Generally, the moisture content of the wood we work with should be between 7-9%.

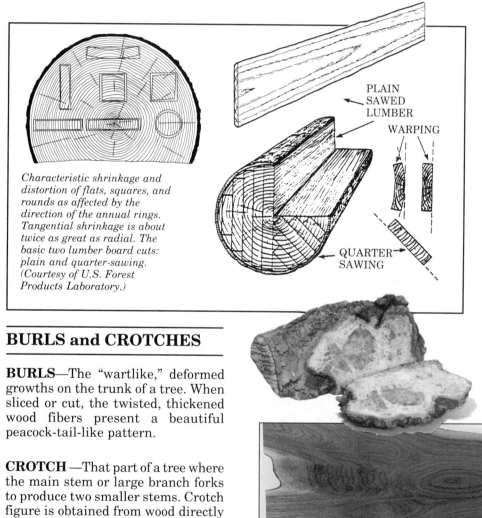

PLAIN SAWED LUMBER

WARPING

QUARTER SAWING

Characteristic shrinkage and distortion of flats, squares, and rounds as affected by the direction of the annual rings. Tangential shrinkage is about twice as great as radial. The basic two lumber board cuts: plain and quarter-sawing. (Courtesy of U.S. Forest Products Laboratory.)

BURLS and CROTCHES

BURLS—The "wartlike," deformed growths on the trunk of a tree. When sliced or cut, the twisted, thickened wood fibers present a beautiful peacock-tail-like pattern.

CROTCH—That part of a tree where the main stem or large branch forks to produce two smaller stems. Crotch figure is obtained from wood directly below the bifurcation and is a result of the stem division.

STRIPPING MATERIALS

Paint and varnish removers, "Strippers" are chemicals that dissolve the finish on the surface to which they have been applied. There is a wide range of these materials available to the consumer, some more effective than others. They are available in a dry powder and liquid form and also a semi-paste type. The liquid variety may be used successfully on flat horizontal surfaces but it is recommended to use the semi-paste type on vertical and irregular surfaces. This type stays where you put it and does not run or drip. The dry powder type is not readily available and has had little impact on those who use stripping materials.

Currently, many new formulations of "strippers" are being offered. Test and select the one that satisfies your endeavors.

Modern marketing now offers many of these liquid strippers in Aerosol containers or spray type applicators.

The use of "Rinse off" strippers is not recommended for use on any quality piece you may work on. The water used to wash off the residue can stain or damage the wood surface.

Be sure to wear proper protective equipment when using these chemicals. Any splash or splatter can and will cause a burning sensation to your skin.

SOLVENTS

LINSEED OIL An oil derived from flaxseed. The raw oil will not dry. Boiled linseed oil is raw linseed oil that has been heat treated and driers added to accelerate its drying abilities. Do not attempt to make your own boiled linseed oil by boiling raw linseed oil. Purchase the boiled linseed oil ready made.

MINERAL SPIRITS Are distilled from petroleum and consist mostly of aliphatic hydrocarbons, hexane, heptane and octane. The composition will vary according to the source of the crude oil and the manufacturer. Chemical analysis has revealed as many as 100 separate compounds in some samples. Mineral spirits are less toxic than most solvents, but the vapors can cause skin irritation, respiratory irritation and depression.

TURPENTINE Is produced by the steam distillation of pine gum. Pine gum contains about 68% solid rosin, 20% turpentine and 12% water. Turpentine has a strong characteristic odor, and is more chemically reactive. Its vapors can cause dizziness and respiratory irritation. It can be a strong skin irritant and cause severe allergic reaction upon repeated contact.

LACQUER THINNER A compound of many coal-tar derivatives such as toluene, benzene and xylene, carefully chosen and balanced to produce a smooth even film when mixed with nitrocellulose. Extremely volatile and flammable.

DENATURED ALCOHOL Ethyl alcohol made unfit for drinking by the addition of a substance such as methanol. It is the solvent used to dissolve shellac gum and also used as a cleaning agent for specific applications.

15

TYPES OF ADHESIVES and THEIR PROPERTIES

When pieces of wood that are to be joined have been properly prepared, the next consideration should be the selection of an adhesive or glue that will assure you of a bond that will hold the pieces together under all conditions that the joined assembly will be subjected to. Will the bonded joint be subjected to heat, cold or moisture? How much stress will the joint be subjected to? What will be the color of the adhesive or glue line when the glue is dry? Is the adhesive toxic? With answers to these questions, you can now select a glue that will bond the pieces together properly, according to your specifications. Some glues or adhesives are purchased ready-to-use; others may have to be mixed. In all cases, carefully read the manufacturer's instructions printed on the container; this will assure you of a better and more professional job.

The following paragraphs will give a brief, general description of five of the most common types of glues used today:

Liquid hide glue.

Hide glue cakes.

Hide glue flakes.

Boiler for heating liquid hide glue.

PVA

LIQUID HIDE GLUE

Perhaps this is the oldest type of wood glue still in use. It is manufactured from bones, hides and tendons of animals. It is available in ready-to-use form, or in flakes or a solid cake. The flakes or cake must be soaked in water for a specified time, then heated in a double boiler for a specified time under specified conditions. The glue is applied hot. It is an excellent adhesive for interior furniture construction and repairs. The glue sets or hardens in about 2-3 hours.

Bonding with liquid hide glue requires the clamping of the assembly. The glue dries to a light brown or amber color. Hide glues are not affected by most finishes or finishing solvents; hide glues do not "load" up sanding belts. The biggest disadvantage of hide glues is that they have poor resistance to water and moisture, and the temperature of the glue mixture is most important in their use.

POLYVINYL ACETATE (PVA or WHITE GLUE)   A most popular, general all around useful adhesive, it is a professional quality woodworking glue and also a good general household adhesive. Clamping of joined assemblies with moderate pressure is recommended. Glue sets in about 1 hour and full strength is obtained in about 24 hours. PVA is non-toxic and dries to a translucent color. A joint bonded with white glue will withstand moderate

16

stress. Do not use a white glue if a waterproof joint is required, and do not use on metal as it will cause corrosion.

ALIPHATIC RESIN (YELLOW) GLUE This is perhaps the fastest growing and most popular of all modern adhesives. Aliphatic resin glues are stronger than white glues. They set in 24 hours. Bonded, assembled pieces provide good resistance to heat and are not greatly affected by varnish, lacquer or paint. Aliphatic glues dry to a very light translucent amber color. They may be pre-colored or stained using water soluble dyes. Aliphatics are water resistant but not waterproof, and are easily sandable.

PLASTIC RESIN (UREA-FORMALDEHYDE) GLUE This glue comes in a powder form and is mixed with water to the consistency of heavy cream. Strong, firm clamping pressure must be applied and maintained for at least 12-16 hours. Plastic Resin glues are toxic. They are water resistant but not waterproof. They may stain acid-content woods, and should not be used on oily type woods. This glue provides a very strong but brittle bond if the joint fits poorly. Plastic resin glues must be used at temperatures 70° or above. They are highly resistant to rot and mold.

RESORCINOL GLUE This glue is supplied in a two part container. One container has the liquid resin and the second container has the powdered hardener. For best results, be sure to follow the manufacturer's mixing instructions on the container. Resorcinol glues are toxic and leave a dark glue line. This type of glue is completely waterproof and excellent for exterior work and boats. Resorcinol glues provide an extremely strong bond and the assembly should be firmly clamped for 16-24 hours. Resorcinol glues must be used at temperatures of 70° and above.

There are many other types of adhesives available. Some have general applications while others have specific applications. The following charts will allow you to rapidly evaluate what type of glue or adhesive you should use for your project.

COMPARISON OF TYPICAL READY-USE ADHESIVES			
	Aliphatic Resin Glue	**Polyvinyl Acetate Glue**	**Liquid Hide Glue**
Appearance	Cream	Clear white	Clear amber
Viscosity (poises at 83°F)	30-35	30-35	45-55
pH*	4.5-5.0	4.5-5.0	7.0
Speed of Set	Very fast	Fast	Slow
Strength (ASTM** Test)	All three easily exceed government specifications of 2800 pounds per square inch on hard maple. On basis of percent wood failure, aliphatic best, liquid hide next, polyvinyl acetate third.		
Stress Resistance†	Good	Fair	Good
Moisture Resistance	Fair	Fair	Poor
Heat Resistance	Good	Poor	Excellent
Solvent Resistance‡	Good	Poor	Good
Gap Filling	Fair	Fair	Fair
Wet Tack	High	None	High
Working Temperature	45°-110°F	60°-90°F	70°-90°F
Film Clarity	Translucent but not clear	Very Clear	Clear but amber
Film Flexibility	Moderate	Flexible	Brittle
Sandability	Good	Fair (will soften)	Excellent
Storage (shelf life)	Excellent	Excellent	Good

*pH—glues with a pH of less than 6 are considered acidic and thus could stain acid woods such as cedar, walnut, oak, cherry and mahogany.
**ASTM—American Society of Testing Materials.
†Stress resistance (cold flow)—refers to the tendency of a product to give way under constant pressure.
‡Solvent resistance—ability of finishing materials such as varnishes, lacquers and stains to take over a glued joint.

COMPARISON of TYPICAL WATER-MIXED and TWO-PART ADHESIVES		
	Plastic Resin	**Resorcinol**
Appearance	Tan	Dark reddish brown
Viscosity	25-35,000 cps	30-40,000 cps
Speed of Set	Slow	Medium
Strength (ASTM Test)	2,800 plus psi	2,800 plus psi
Stress Resistance	Good	Good
Moisture Resistance	Good	Waterproof
Heat Resistance	Good	Good
Solvent Resistance	Good	Good
Gap Filling Ability	Fair	Fair
Wet Tack	Poor	Poor
Working Temperature	70°-100°F	70°-120°F
Film Clarity	Opaque	Opaque
Film Flexibility	Brittle	Brittle
Sandability	Good	Good
Storage (shelf life)	1 year	1 year

PROPERTIES OF WOODS
WIDELY USED FOR GLUED
PRODUCTS

HARDWOODS			
Group1	**Group 2**	**Group 3**	**Group 4**
(Glue very easily with different glues under wide range of gluing conditions.)	(Glue well with different glues under a moderately wide range of gluing conditions.)	(Glue satisfactorily under well-controlled gluing conditions.)	(Require very close control of gluing conditions or special treatment to obtain best results.)

Aspen
Chestnut,
  American
Cottonwood
Willow, black
Yellow poplar
	Alder, red
	Basswood
	Butternut
	Elm:
	  American
	  Rock
	Hackberry
	Magnolia
	Mahogany
	Sweetgum
		Ash, white
		Cherry, black
		Dogwood
		Maple, soft
		Oak:
		  Red
		  White
		Pecan
		Sycamore
		Tupelo:
		  Black
		  Water
		Walnut, black
			Beech, American
			Birch, sweet and
			  yellow
			Hickory
			Maple, hard
			Osage-orange
			Persimmon

SOFTWOODS		
Group1	**Group 2**	**Group 3**

Baldcypress
Cedar, western red
Fir, white
Larch, western
Redwood
Spruce, Sitka
	Cedar, eastern red
	Douglas-fir
	Hemlock, western
	Pine:
	  Eastern white
	  Southern yellow
	  Ponderosa
		Cedar, Alaska

Courtesy of Franklin Chemical Industries

ABRASIVES

One material every woodworker uses is commonly called "Sandpaper." However, that word is a misnomer when referring to the abrasive materials we use today. The modern abrasives are separated into two classes, natural and synthetic. The natural abrasives are Flint, Garnet and Emery. In the synthetic class we find Boron Carbide, Silicon Carbide, and Aluminum Oxide.

These materials are ground, screened, and graded according to grit size, then applied to a variety of backing materials using a variety of adhesives. They are available as flat sheets, discs, belts, cords, tapes and other related applications.

The chart on page 66 will guide you in the selection of the proper abrasive for your immediate requirements.

Mineral Garnet, Powdered Garnet, and Garnet Paper

Silicon Carbide Finishing Paper   *Drum Sanders*

Cord and Tapes

Sand-O-Flex

TACK CLOTHS

The use of a Tack cloth is a must for the ultimate finish. After any sanding operation, it removes any loose dirt, sanding particles or lint that is too small to see but can leave a blemish on the finished surface. Tack cloths can be purchased ready made, or you can make your own as follows.

Secure a piece of clean, washed cheesecloth, dip into container of clean turpentine, wring out thoroughly, then add about a tablespoon of varnish or polyurethane to the cloth and knead thoroughly. This is your tack cloth. When not using the tack cloth, keep it in a tightly closed glass jar with a few tablespoons of turpentine in the bottom. This will keep the cloth flexible and usable.

COLORING MATERIALS

JAPAN COLORS Highly concentrated Colored pigments mixed with a vehicle that is compatible with either **oil** or **lacquer** based products. It is available as a thick paste and must be thinned before mixing it with other products. Do **NOT** use with water based materials such as Latex.

OIL COLORS A heavy bodied combination of concentrated pigments mixed with linseed oil and other oils. It is used to tint **oil**-based products. It should be thinned with the recommended vehicle before mixing with the finish.

UNIVERSAL COLORS Tinting colors that are compatible with **oil** or **water** based products such as Acrylics and Alkyds. Their liquid consistency makes them easier to mix than the paste-type tinting colors. Even though they are called universal, they **may not** be compatible with **some** lacquers, epoxies, or catalyzed finishes. Check for compatibility before using these products.

FRESCO COLORS Finely ground dry pigments used for coloring, mixing and tinting. May be used with enamels, fillers, shellac, lacquers. Very useful for touch-up work. Their extreme fineness allows them to mix readily with a wide variety of mediums to produce a smooth opaque finish.

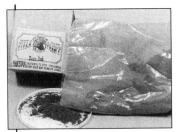

MILK PAINT   A combination of milk products, mineral fillers and natural earth colored pigments. Available in powder form which is mixed with water at time of use. Apply to a raw wood surface or one that has been thoroughly sanded.

Milk paint when completely dry is nontoxic. It has a shelf life of one year when in the powder form and stored in an air-tight container.

STAINS

There are three basic types of stain, classified by the type of solvent or vehicle used to dissolve or suspend the dye material. They are: Water Stain, NGR Stain and Oil Stain.

WATER STAINS   These give the clearest and most transparent results of all the stains. They penetrate deep into the wood and have the least tendency to fade and "bleed" into subsequent coats. They are the lowest in cost.

Water stains are available as water soluble aniline dye powders. One ounce of the stain powder is dissolved in one quart of warm water. Due to the high mineral content of water in some areas, it is recommended that distilled water be used for mixing water stains. This eliminates the possibility of any minerals in the water contaminating or altering the color of the stain.

The strength, hue or "depth" of the color can be varied by varying the amount of water used; more water for a lighter color, less water for stronger or darker color. Water stains should be mixed and stored in an earthenware or glass container.

During the process of milling or planing the wood surface, the wood fibres are flattened down. Since water is the vehicle for the stain, the water will raise the grain of the wood surface unless the surface is properly prepared as follows: Use a sponge **dampened** with water and sponge the entire surface you are going to stain. This will raise the wood fibers. Let dry, then use fine sandpaper (#240) and sand off these raised fibers, sanding in the direction of the grain. Apply the stain and the surface will remain smooth since you have already sanded off the fibers that would have stood up.

The use of a spray gun is the ideal method to apply water stains. A brush, rag or sponge may also be used, but it will require practice and patience to produce a uniform even coat of water stain on a wood surface. Allow the stained surface to dry at least 24 hours prior to any further finishing applications.

NGR (Non Grain Raising) STAINS These are aniline dyes soluble in nonaqueous solvents such as alcohol, glycol, toluol, acetone and other ketones.

They produce bright, transparent colors which are, light fast, but not as light fast as water stains. Due to the volatile solvents, they dry rapidly, making the application by brush a difficult task. They are ready for subsequent finishing procedures in 4-6 hours.

If possible, a spray gun should be used to apply NGR stain. A brush, rag or sponge can be used, but it will take practice and patience to produce a good uniform job of staining with NGR stain.

Non-Grain Raising Stains are non-bleeding and permanent in color. They offer all the fine qualities of Water Stains but avoid the difficulties of raising the grain. These stains should be applied by either spraying or wiping on with a rag. Repeated thin applications are recommended to minimize overlaps. One color can be applied directly over another but too dark a color must be bleached out.

Non-Grain raising stains are recommended for hardwood. These stains can be intermixed to give a wider variety of color, or a special Clear Thinner can be added to reduce the intensity or depth of a color. The dyes used in these stains are strong in color and it is recommended that you protect your skin and clothes during application. No sanding is required after application, and after four hours you may proceed with your next finishing operation. Always test the stain on a scrap of the same type of wood you are going to ultimately stain, or on some inconspicuous spot of the surface you are going to stain.

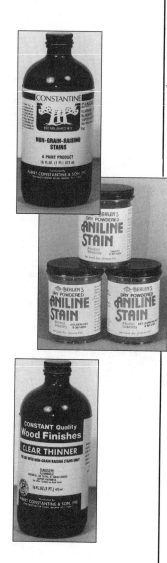

Pigmented Oil Stain.

OIL STAINS The ideal stain for use by an amateur. They are dyes or colorants mixed into a vehicle such as linseed oil or tung oil, to which has been added driers and thinners such as turpentine. Oil stains are not as transparent as water or NGR stains and they are prone to fade when exposed to direct sunlight.

There are **two** basic types of oil stain, **Penetrating** Oil Stain and **Pigmented** Oil stain. In the Penetrating Oil stain, the dyes used are in liquid form and mixed into the solvent or vehicle. This type of stain penetrates deep into the wood. The Pigmented Oil stains, often referred to as "wiping stains," are made from dyes ground and mixed into the vehicle. These small dye particles will settle to the bottom of the container; hence it is necessary, when using a Pigmented Type stain, that the contents be thoroughly mixed prior to and during use.

Due to natural slow drying time of the solvents, you have ample time to adjust and blend the color during application. Oil stains are easily applied with a spray gun, brush, rag or sponge. Excess stain can be wiped off with a cloth before it sets. The stain should be allowed 24 hours to dry prior to any further finishing procedures.

Penetrating Oil stains.

24

FILLING MATERIALS

PASTE WOOD FILLER

A compound of Silex, boiled linseed oil, turpentine and driers, used to level the surface of wood by filling up the open pores and grain. Often a pigmented coloring medium such as Japan colors is added. These colored paste wood fillers also help to accent the grain of the wood.

Ready made paste wood fillers are available in about seven different colors: Red Mahogany, Brown Mahogany, Oak, Walnut, White, Black and Neutral.

 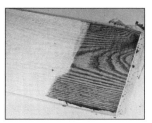 

WOOD FILLER

A material used to fill up voids or missing pieces of wood, it may be purchased under trade names such as Carpenter's Wood Filler, Plastic Wood, Wood Dough, Powdered Wood Compound, Putty, etc. These materials, when dry, are hard, and may be drilled, sanded, or carved. They are ideal for filling up holes and spaces that will be painted over. Be aware that some fillers will "shrink" when dry and do not mix more filler than you expect to use immediately.

When dry, most wood fillers are very difficult to stain and finish so they are not noticeable. Some of these fillers are advertised as stainable. Carefully read the instructions on the container. Often this staining must be done while the material is "Wet" and pliable and the use of a compatible stain is mandatory. Some manufacturers offer "Fillers" that are already stained or colored.

The areas where fillers are used should be clean and lightly sanded. The filler is applied a little at a time working from the bottom of the hole, up. The top surface of the newly applied filler should be slightly above the surrounding area allowing the filler, when dry, to be sanded level with the surrounding area.

Do not confuse these wood fillers with "Paste Wood Filler." They are NOT the same!

SHELLAC STICKS These sticks of hard shellac are available in a wide variety of colors, some opaque and some transparent. The opaque sticks are used to fill up indentations where both the stain and finish are missing. The transparent sticks can be used to fill up shallow indentations where the original shade of the finish remains.

A small, heated spatula is used to transfer the molten shellac into the hole or indentation, which when cool can be leveled with the surrounding surface, stained and finished. An Electric "Burn In" knife may be used in place of the alcohol lamp and spatula.

FIL-STICKS or PUTTY STICKS Sticks made up of a combination of waxes and other pliable finishing materials. Available in a wide variety of colors and ideal for filling minor scratches and holes.

FINISHING MATERIALS

Finishing is the process of coating a surface to decorate, and/or protect the surface from unwanted exterior sources. The following is a brief overview of some of the most common and modern materials used as finishes today.

Finishes can be placed into two broad categories:

SURFACE FINISH

Dries by evaporation or oxidation.

PENETRATING FINISH

Penetrates into the wood.

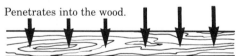

Protects from within the wood surface.

Polymerizes (solidifies).

SURFACE FINISHES

Finishes that are applied to a surface and dry by either oxidation or evaporation, leaving a protective film on the surface. The vehicle is solvent or liquid.

PENETRATING FINISHES

Finishes that are applied to a surface, penetrate the surface and harden or polymerize from below the surface up to the top.

There are many variations or combinations of these two types of finishing materials.

SHELLAC As it is known today, refers to all forms of purified lac. Lac is the hardened resinous secretion of a tiny scale insect, found predominantly in India. It is a parasitic animal which uses certain trees and shrubs for its host, and secretes a resin as a protective shell for its larva. Lac has been known in India for two to three thousand years. Refined, ready mixed shellac, offered for sale to the public, has a shelf life of six months. The date of manufacture is usually marked on the container. The use of old or outdated shellac may result in the material not drying and remaining gummy.

When using shellac, it is often recommended that the shellac be reduced or thinned, using the terms "Lb.Cut." The following chart is provided to assist you in making the proper recommended shellac mixture.

Shellac in its original raw form.

L to R: Shellac flakes, powdered shellac, bleached shellac.

THINNING CHART	
To reduce shellac, follow these thinning ratios:	
To Convert:	**Add This Much Alcohol:**
5 lb. cut to 3 lb. cut	7/8 pt. to 1 qt. shellac
5 lb. cut to 2 lb. cut	1 qt. to 1 qt. shellac
5 lb. cut to 1 lb. cut	2/3 gal. to 1 qt. shellac
4 lb. cut to 3 lb. cut	1/2 pt. to 1 qt. shellac
4 lb. cut to 2 lb. cut	3/4 qt. to 1 qt. shellac
4 lb. cut to 1 lb. cut	2 qts. to 1 qt. shellac
3 lb. cut to 2 lb. cut	3/4 pt. to 1 qt. shellac
3 lb. cut to 1 lb. cut	3 pts. to 1 qt. shellac

Why Shellac Is Thinner in Warm Weather
Alcohol expands under heat and also becomes a better solvent for shellac. This tends to reduce the viscosity of shellac. A 4 lb. shellac, for example, will be noticeably thinner in body on a 90° F. (32° C.) day than at 65° F. (18° C.)
This explains why in the summertime you will occasionally have a complaint that shellac is "too thin." Despite the thinner feel, however, there are just as much solids as always. The shellac will still provide the same sealing, hiding and film characteristics as it does in cooler weather.

For thinning shellac, use a good proprietary denatured alcohol, one that is recommended on the label of the container. The use of substitute spirit thinners may impair the film properties of the shellac.

Shellac as a total or final finish, is seldom used today except in the purist form of "French Polishing." Shellac in the "cut form" is often used as a sealer or a Wash Coat, but should NOT be used under a polyurethane finish.

LACQUER As far back as 500 B.C., Chinese and Japanese craftsmen were using a liquid exudation (sap) from certain lac trees to cover or finish the surfaces of their wooden possessions. This material provided a surface that was covered and protected with a clear hard film. This early form of lacquer dried by oxidation while our modern type lacquers dry by evaporation. Over the years the base and solvents for our modern lacquers were developed. In 1855 Nitrocellulose lacquer was offered for commercial use and has been the base for most lacquers since that time.

Research chemists continued to experiment with the base formula, adding materials to color the lacquer, adjust the sheen, make the lacquer more flexible, increase the resistance to abrasion, certain chemicals and liquids and also increase the resistance of lacquer to variations of temperature.

Lacquer has come to denote a finishing material that dries very rapidly by evaporation, leaving a protective film after the volatile solvent has evaporated. Because of the low solid content of lacquer, it requires the application of multiple coats to produce a film thickness equal to that of an oleoresinous or synthetic finish.

If you use a lacquer thinner, it is recommended that it be from the same manufacturer of the lacquer.

If lacquer is used as a finish, do NOT apply it over any other type of finish such as varnish or polyurethane.

28

ADVANTAGES OF LACQUER	DISADVANTAGES OF LACQUER
1 Dries rapidly, dust free in about 15 minutes.	**1** Lacquer is difficult to apply with a brush. Good spray equipment should be used to apply lacquer.
2 It is relatively easy to repair a marred or damaged spot in a lacquered surface.	**2** If lacquer is to be applied with a brush, it should be purchased as a "Brushing Lacquer." Alternatively, a "Lacquer Retarder" (which slows down the drying time) may be added to a Spraying Lacquer along with additional thinner to make it "brushable."
3 When dry, a lacquered surface is hard and relatively resistant to abrasion, water, heat and mild alkaloids and acids.	
4 Lacquer is the clearest of all finishes with no color of its own to alter the color of the stain or wood surface to which it is applied.	

PADDING LACQUER A special formulation of lacquer applied with cloth pads, used as a substitute for Shellac in the modern version of French Polishing. Often used in combination with blending powders, for spot repairs on a finished surface. May be applied over shellac, lacquer, varnish, polyurethane and penetrating type finishes.

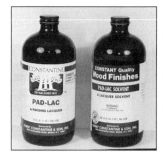

POLYURETHANE The modern synthetic type varnish. In 1957 research produced and offered the craftsman a "varnish" that was made entirely from chemicals. It is by far the most versatile finish offered today.

Polyurethane was actually discovered in 1848 by a German chemist who reactivated isocyanate with alcohol. After 1957 the urethanes were modified to eliminate the poisonous isocyanate by using vegetable oils and became known as "polyurethane." There are many variations of this formulation today and they are often referred to as varnish.

ADVANTAGES OF POLYURETHANE	DISADVANTAGES OF POLYURETHANE

ADVANTAGES OF POLYURETHANE

1 Polyurethanes dry quickly and "may" be recoated in 4 hours.

2 Polyurethane may be applied with a brush, spray, or plastic applicator and is easily applied over **most** other type finishes that have been properly prepared.

3 Polyurethane finishes are extremely mar and abrasion resistant. This, combined with their high resistance to stains, water and alcohol, makes them the ideal finish for floors and furniture that will be subjected to wear.

4 Polyurethane may be tinted using Japan or Oil colorants.

5 Polyurethane finishes are more durable than natural varnish finishes

6 In its pigmented form, "Enamel," polyurethane has a high color retention property.

DISADVANTAGES OF POLYURETHANE

1 Polyurethane should not be applied over sealers such as shellac or lacquer, unless they have been specially formulated for polyurethanes.

2 Polyurethanes do NOT bond chemically to another surface. Such a surface must provide a good mechanical bond which can be achieved by a thorough sanding job.

VARNISH A clear finishing material made from resin, drying oil and thinners. The use of varnish can be documented back to the early Egyptians who combined the resins from Flax together with Olive Oil or Cedar Oil to form their varnish. Over the years, through research and necessity, the composition of varnish has been altered and refined. Various resins and oils, some natural and some synthetic, are mixed, heated, distilled and combined with other chemicals to produce a specific varnish with specific characteristics.

True, pure Varnish is a scarce item today. It has been replaced almost entirely by Urethanes and Polyurethanes, whose properties equal and exceed those of pure Varnish.

SPAR VARNISH A special varnish (urethane) formulated to resist the potential devastation of exposure to natural elements such as dampness, salt water, plain water, sun, etc. Highly recommended for all exterior varnishing projects.

TUNG OIL VARNISH A thin transparent finish that wipes on easily and penetrates deep into the wood, accenting the true beauty of the wood grain. Apply only to a new wood surface or to an old surface that has been properly prepared. Apply with a soft cloth and rub vigorously into the wood surface. One coat produces a low luster, two coats a satin sheen and three or more coats a higher gloss. It is resistant to moisture, fast drying and durable.

TUNG OIL Also known as "China Wood Oil," it is an oil finish that can be used full strength or thinned with mineral spirits or turpentine. Apply only to new or properly prepared stripped wood. Apply with a cloth. Dries to a hard finish which is resistant to water, acid, alkalis and heat. One coat produces a low luster finish. Subsequent coats increase the gloss luster.

PENETRATING FINISH An oil modified polyurethane. It penetrates deeply into the wood surface enhancing the color and grain of the wood. It is resistant to both water and alcohol. Apply with a brush or soft cloth over bare wood. One to three coats are recommended depending upon the luster desired. Easy to retouch.

Antique Oil Finish by Minwax is a one-step method which provides hand-rubbed beauty in minutes. Simply wipe the surface with a lint-free cloth moistened with the finish. Then, buff evenly when the surface is tacky to the touch. This will give the wood a new healthy glow and a tough layer of protection at the same time. For more severe cases, apply Antique Oil with a fine grade of steel wool, rubbing with the grain.

DANISH OIL FINISH　A Super penetrating, resin-oil, grain revealing finish. It primes, seals, stabilizes and hardens the wood up to 25%. May be applied to a new wood surface or a stripped surface that has been properly prepared. Apply with a brush, cloth, foam applicator, roller or spray. It produces a surface that has a low-luster, hand rubbed look, is resistant to minor scratches and stains and is easy to retouch.

MISCELLANEOUS FINISHING MATERIALS　The finishes listed above are by no means the only ones available today. They are but a few of the most common. Literally thousands are available to the consumer, many being combinations or varied formulations of some of the above materials. Be sure to check the printed instructions on the container prior to use.

Be aware of some relatively new innovations being presented for those who are interested in finishing, such as Gel type stains and finishes, water emulsion lacquers and polyurethanes, catalyzed lacquers, Table Top Resins, rubber base paints, etc.

RUBBING and POLISHING COMPOUNDS

Pumice

Rottenstone

Rubbing Oil

PUMICE STONE Manufactured from natural Lava rock. The rock is pulverized and the particles are separated and graded according to their size; the finest being 4F down through 3F, 2F, and 1F, the coarsest.

ROTTENSTONE A natural substance, namely the product of decomposed siliceous limestone. Rottenstone is a finer abrasive than Pumice stone and is available in only one grit.

Both Pumice Stone and Rottenstone are provided to the consumer as fine, dry powders. For ease of application and storage, it is often advantageous to secure two or more inexpensive shakers and place these abrasive powders in the shakers, being sure to mark the shakers as to their contents. When not in use, these materials should be placed in a dry storage area.

Both Pumice Stone and Rottenstone are usually used with a lubricant, water or rubbing oil, depending upon the luster desired and the surface material being rubbed.

POLISHING COMPOUND A semi-paste type material composed of very fine abrasives such as Jewelers Rouge or Tripoli combined with oils and waxes. It may be used for the ultimate rubbing of a finished surface producing a mirror like finish and may also be used to remove superficial scratches or dirt from a finished surface.

Repairing and Preparing for the Finish

- WASHING and CLEANING
- STRIPPING
- BLEACHING
- PHYSICAL REPAIRS
- DISASSEMBLY
- JOINING WOOD SURFACES
- TYPES OF JOINTS
- REPLACING BROKEN RUNGS or SPREADERS
- TIGHTENING LOOSE JOINTS
- VENEER REPAIRS
- REMOVING DENTS IN RAW WOOD
- FILLING CRACKS and HOLES
- PREPARATION OF WOOD SURFACE FOR FINISHING

To achieve professional, high quality results in the restoration and finishing of your project, a definite sequence should be followed. Prior to the application of any finishing material, either stain or a finish, the surface must be properly prepared and the following steps should be considered if applicable.

To achieve the ultimate durable finish and permit full use of the piece of furniture again, the piece should be carefully cleaned to remove all accumulated dirt and grime. This will expose the true condition of the surface and allow you to evaluate the condition of the remaining finish. The decision to remove the remaining old finish or repair the old finish will be up to you.

If stripping off the old finish becomes your option, when that is completed, bleaching would be the next step if required. This should be followed by making any necessary physical repairs, whether it be replacing broken or missing pieces, repairing broken or loose joints, making repairs to broken or buckled veneered surfaces. Any dents or cracks should be filled and, finally, the entire surface properly sanded to accept whatever finishing material you elect to use.

WASHING and CLEANING

Once you have acquired your old "masterpiece," your first consideration should be to evaluate the true condition of the finish which over the years may have acquired a build up of dust, smoke, grease, dirt and grime that can not be removed by normal dusting. It is therefore necessary to remove this layer of camouflage to expose the true condition of the finish, color and original grain. This will allow you to consider what repairs and further finishing steps, if any, are necessary.

The recommended first step is to simply wash the piece very carefully.

Often this will be all that is necessary to expose the true finish, color and the original grain of the wood, providing the piece still has some finish on it. A solution of Murphy's Oil soap or a small quantity of a mild detergent dissolved in warm water is all that is necessary to clean the piece. Cotton fabric such as cheesecloth, old toweling or T-shirts make excellent cloths to be used for the cleaning process. Do not use synthetic type fabrics.

Dip a piece of the cloth in your cleaning solution and wring out about half of the solution. Work quickly and gently, washing a small area at a time, wiping in the predominant grain direction. Remove the cleaning solution from this area by wiping with a cloth wrung out in clear, clean, warm, water. Follow this by drying with yet another clean, soft cloth.

Continue this cleaning process, working small areas at a time until you have covered the entire piece. Finally, completely dry the entire surface with a soft, clean, dry cloth. It may be necessary to repeat this washing and drying operation several times to remove all the residual dirt and to expose the true, real finish.

When the finished surface is truly exposed, carefully examine the finish. If you determine that the finish is not in need of repairs or refinishing, you may elect to apply a coat of wax or a good furniture polish (OZ polish) and once again place the piece of furniture back into active use, and proudly display it.

STRIPPING

After all the dirt and grime has been removed, carefully examine the exposed finish. If you determine that the surface finish is beyond restoration, the only alternative is to completely remove all the old finish. This "stripping" process is perhaps the least glamorous and most messy part of the entire refinishing process.

There are two ways to remove an old or unwanted finish from a surface. First, mechanically, using scrapers or sanding, and in either case you risk removing some of the patina and wood itself. Secondly, and less risky to the patina and wood surface, is by the use of chemicals. The use of modern "Heat Guns" is ***not*** recommended for removing old finishes from furniture

Commercial Stripping enterprises are available for those who elect not to do this part of the restoration themselves. These establishments often resort to dip stripping where the entire piece of furniture is tossed into a tank of stripping chemicals, subjecting all joints and veneered surfaces to loosening or coming apart. Unless you intend to completely rebuild and reglue the entire piece, it is strongly recommended that you "hand strip" the piece, especially if the piece has any value as an authentic antique.

Paint and Varnish removers (strippers) are available in liquid, semi-paste, and paste types. For the individual who elects to do the stripping themselves, the semi-paste type remover is highly recommended.

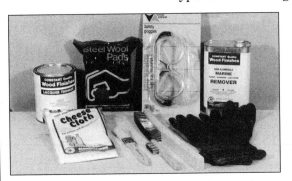

Before starting the stripping process itself, remove any hardware: hinges, knobs, handles and locks. These can be cleaned, polished and a protective coat applied later on. Remove drawers and doors and work on them as separate pieces. Prepare the area where you are going to work. Cover the area with old newspapers, have adequate ventilation, avoid all excessive heat sources such as pilot lights, sparks and flames.

Assemble everything you will need: stripper and/or paint remover, lacquer thinner, a clean, empty can, brush, scrapers, brass brushes, goggles, cheese cloth, gloves.

Above all, protect yourself. Wear old clothes, an old long sleeved shirt, apron, heavy neoprene gloves, goggles and or a protective face shield.

Pour a quantity of your selected remover into an empty container (do not use plastic). Use a clean, old or inexpensive stiff fiber bristle brush. Do not use a brush with synthetic type bristles; the remover may dissolve them. Flow on the remover liberally to an area about 1-2 sq. ft. being careful not to splash or spill any of the remover, especially on yourself.

Apply remover in the general direction of the grain. Do not brush back and forth. After about ten to twenty minutes, when the surface becomes soft, you should see the surface start to wrinkle or blister. It is now ready to be "stripped" or removed. On flat surfaces, use a blunt putty knife whose sharp corners have been filed off and slightly rounded to lessen the possibility of gouging the wood. Scrape and *lift* off the loosened old finish, don't smear it around. Dispose of this "gunk" into another old empty can.

On irregular or carved surfaces, use a brass bristle brush, ***not*** steel bristles (they are too stiff and will add more grain to the surface). Coarse steel wool or an abrasive sponge will also effectively remove the old "sludge" from irregular surfaces. Carefully rub or brush off the loose old finish. Use small fine scrapers or picks to remove all the old loose finish from all the intricate carvings or corners, being careful not to scratch or damage the wood surface.

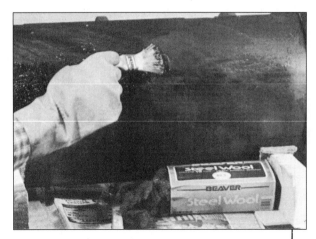

Using a stiff fiber bristle brush, brush remover on in direction of grain.

Let paint remover "Soak in" 10-30 minutes, or until surface shows wrinkles or blisters.

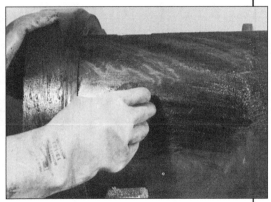

Remove the "gunk" with a scraper or coarse steel wool.

If the old finish is not completely removed the first time, it may be necessary to repeat the process. If you find the surface hard or stubborn to remove, build a tent over the piece using a sheet of heavy plastic; this will concentrate all the chemical vapors within the tent rather then allowing some of the vapors and chemical action to dissipate into the air. Do NOT allow the plastic tent to touch any part of the surface that has been covered with the stripping material.

Use brass bristle brush (not steel) on irregular or carved surfaces.

Strippers or Removers will only remove surface materials. They will **NOT** remove burn marks or stains in the wood surface.

Some of the modern type finishes such as Table Top Resins and Epoxy materials are impossible to remove with chemicals. These finishes have to be removed mechanically by sanding or scraping.

REMOVING MILK or REFRACTORY PAINT

Sometimes on old antique surfaces, you may find the painted surface difficult to remove. If so, it was probably made from milk or casein and is often referred to as Milk Paint. If you encounter this problem, try the following:

Apply a heavy coat of remover to the surface, allow it to stand for 10-15 minutes and, while the remover is still wet, scrub the surface with a brass bristle brush or pads of coarse steel wool. Keep the steel wool or brush clean. Repeat this process and follow by rubbing with a pad of coarse steel wool dipped in denatured alcohol. Full strength ammonia may be substituted for the denatured alcohol, but it can and may cause a stain in the wood, in addition to causing breathing problems.

Very often, it is impossible to remove all traces of this type of obstinate paint using chemicals. Make a test first in a very small inconspicuous area. Sanding may be your only recourse for removal of old Milk Paint.

NEUTRALIZING

All paint and varnish removers are chemical compounds and any residue that remains on the wood surface could react with any stain or finish that you may ultimately apply to the surface. It is recommended that, after all the old finish is removed, the surface be neutralized to prevent any possible reaction you might encounter with any residual amount of the paint and varnish remover that may be left in or on the surface of the wood. This neutralization can easily be accomplished by wiping the entire surface with a clean cloth dipped into lacquer thinner or denatured alcohol. These materials will evaporate in a very short period of time and you can proceed to work on your project being assured that you have eliminated one potential subsequent aggravation.

After stripping, neutralize the surface with denatured alcohol or lacquer thinner.

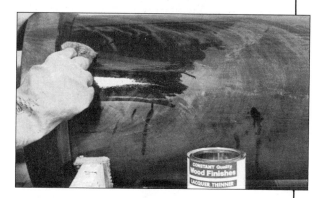

Wipe entire surface with a cloth or fine steel wool dipped into lacquer thinner or denatured alcohol.

TIPS & TECHNIQUES

1 When stripping and washing vertical surfaces, always start from the bottom and work up. This will help to eliminate unsightly "run" marks.

2 If you do get a burn from splashed stripping material, wash the affected area with clean water. If the material gets into your eyes, head for the nearest hospital.

3 If the area where you have applied the stripper becomes dry, simply recoat with fresh remover.

4 Brass bristle brushes are easily cleaned with soap and water followed by combing with a metal dog type comb to assist in removing all the "gunk" from the brush.

5 When you are completely finished with your stripping, **neutralize** the surface with lacquer thinner or denatured alcohol.

41

BLEACHING

Once all the old finish has been completely removed from the wood surface, any darkened or discolored spots or areas can usually be lightened or removed by the proper use of Bleaching agents.

There are three types of bleaches readily available for use by the do-it-yourselfer:

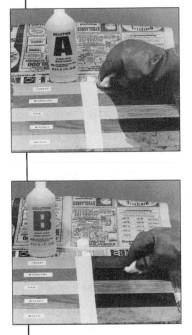

1. **Common household laundry bleach,** usually effective on woods such as maple, walnut and gumwood. Its use is suggested to be confined to spot work only. With a sponge, brush or cloth, apply a liberal amount of the bleach over the spot or surface of the wood. After 15-20 minutes, wash the surface with clear water. This process should be repeated until the desired color has been attained. Neutralize by washing with a 50% solution of **white** vinegar and warm water. Rinse with clear water and allow the wood surface to dry thoroughly at least 48 hours prior to any further work on that surface.

2. **Prepared commercial bleach**, usually available as a two part solution from large paint stores and selected wood product establishments. The materials can be intermixed or applied separately, carefully following the manufacturer's instructions. Use either a sponge, brush or cloth to apply the bleach. Allow to dry until the desired color is achieved. Repeat the process if necessary. Neutralize with a solution of 50% **white** vinegar and warm water. Rinse surface and allow to dry thoroughly.

3. Oxalic Acid, available at large paint stores in a crystal or powder form which is mixed with hot water. Mix 10-16 ounces of the crystals or powder to a gallon of hot water; this makes an ideal solution. Apply liberally with sponge, brush or cloth. After 20-30 minutes wipe off solution with a damp cloth. Repeat process until desired color is obtained. Neutralize with a solution of 3 ounces of borax to 1 gallon of water, rinse thoroughly and allow entire surface to dry 48 hours.

It is most important that, after you have bleached the wood to the color you want, you neutralize the surface you have bleached. This neutralizing process helps to prevent any possible chemical reactions with other materials that you may put on the surface such as filler, stains and finishes.

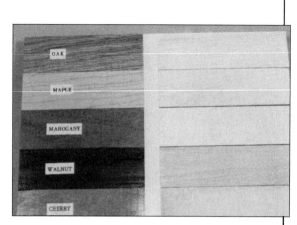

Unbleached   *Bleached*

Never mix bleaching agents and ammonia together.
This produces poisonous fumes!!!

TIPS & TECHNIQUES

1 Wear adequate protective clothing, gloves and eye shield.

2 Bleach in a room with an even temperature. Avoid direct sunlight and drafts.

3 Avoid use of metal containers for bleaching materials.

4 Do not use animal bristle brushes to apply bleaching materials.

5 When finished bleaching—Neutralize.

PHYSICAL REPAIRS

Today's high cost of furniture replacement, whether old or new, makes it imperative for the home owner to seriously consider keeping what he or she owns and upgrading it when it becomes necessary.

As you look around your home, you will probably find a few pieces of furniture in need of some physical repairs. The most likely candidates for attention are chairs, followed by tables and cabinets. Most problems may be unique for you, but in reality they are universal. Many ailments, and their cures, are common to chairs, tables and cabinets, and the following discussions and techniques are applicable to most all wood assemblies.

Every assembly of wood pieces incorporates some type of joint. There are relatively few types of joints but there are hundreds of variations of these basic joints. Veneer has been used for centuries to enhance the beauty of wood surfaces, and this veneer is often in need of repair or replacement.

Some of the most commonly diagnosed ailments, combined with suggested techniques and cures, will be discussed in detail in the following paragraphs.

It is necessary to evaluate how extensive you intend to make any structural or physical repairs. Is the piece of furniture ultimately to be placed into full use again or will it be just for display? Is the repair to be permanent or just cosmetic?

Every repair is unique and the course to be followed should be carefully thought out. What will be the progression of the steps to be followed? What materials will be required? What tools should be used? In addition to these physical requirements, you will have to supply a good portion of common sense, ingenuity and patience.

DISASSEMBLY

While you are evaluating what repairs will be necessary, consider how completely you may have to disassemble the piece to effect the necessary repairs.

It will prove to be an advantage that, when pieces are disassembled, the corresponding pieces of the matching joints are marked. This will remove the possibility, during final reassembly, of putting the wrong piece in the wrong hole. Such occurrences can happen, especially when there are multiple pieces similar in shape or size such as chair rungs, spindles in chair backs or head boards of a bed, etc.

All joints incorporate the use of some type of glue. It may be necessary to break this glue bond, but don't break the joint unless it is absolutely necessary. Most glues can be softened with a 50% mixture of *white* vinegar and warm water. Use a pin or a sharp pointed knife and put several small holes in the jointed area, then carefully apply your mixture to this area. The use of a Glue Injector to concentrate the application of the mixture is ideal. Shortly you will find that the old glue has become soft and the bond can be broken without damage to the surrounding wood surface.

Marking pieces of joints for matching.

Glue Injector

If a hammer or mallet is used to assist in the disassembly, be sure to protect the surface that you are going to strike with a piece of felt or other heavy fabric. Direct all blows as close to joint as possible. Don't damage the wood surface.

45

JOINING WOOD SURFACES

There are several ways two pieces of wood can be fastened together, such as nailing, using screws or bolts; however, the ideal way to join wood is with glue, with an adhesive that forms a strong bond between the wood surfaces. A properly glued joint will often be stronger than the wood itself. It will be strong, durable and will look good. The quality of the glue joint depends on several factors, including the type and condition of the wood. The preparation of the wood surface, new or old, is important, and the type of joint that is used will also make a difference. Other considerations are the type of adhesive that is used, the atmospheric condition, the temperature and humidity of the shop or room where the work is being accomplished, the correct amount of pressure, the location of the clamps and the length of time the joint remains under pressure. All these elements are all essential to good joining.

TYPES OF JOINTS

Properly seasoned wood can be joined in a variety of ways. If the wood is going to be finished with a clear, natural finish like oil, stain or varnish, some thought should be given to matching the grain of the different pieces of wood. The grain of the wood is not just decorative; some consideration has to be given to its strength characteristics.

Gluing Narrow Stock to make wide panel.

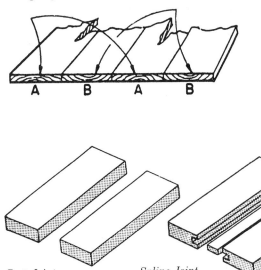

Rings Up   *Rings Down*

Edge to Edge joining is very common and fairly simple. In edge to edge joining, the grain of the pieces of wood is parallel. When a number of pieces are joined edge to edge, the growth rings should be alternated, the direction of the rings should be reversed from one piece to the next. This helps to prevent cupping and twisting.

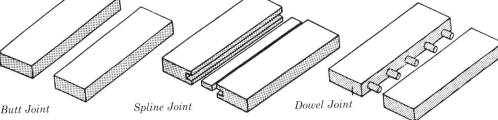

Butt Joint   *Spline Joint*   *Dowel Joint*

End to End joining is more difficult. The end pieces of the wood tend to absorb too much of the glue; also it is hard to get a very smooth surface on the end grain and to get a good bond. It is important that the end surfaces are smooth.

The best way to splice two pieces of wood, to join two ends, is to expose some of the side grain. There are a number of ways to do this, as illustrated. The main thing is to expose some of the side grain and enlarge the amount of surface that is going to be glued.

END TO END JOINTS

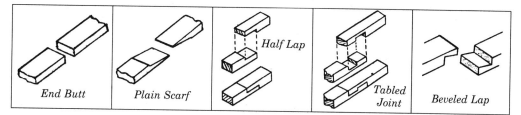

End Butt   Plain Scarf   Half Lap   Tabled Joint   Beveled Lap

End to Side grain joints are very common and useful, but because the wood changes dimensions when there is a change in the moisture content, the joints have to be very well designed and very well prepared. Although there are only a few basic types of end to side grain joints, there are probably over a hundred variations of these basic joints. Some are very complex, some are very delicate, some are extremely strong. A few examples are illustrated below.

END TO SIDE BUTT JOINTS

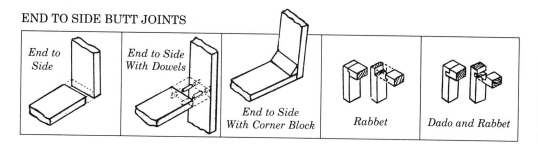

End to Side   End to Side With Dowels   End to Side With Corner Block   Rabbet   Dado and Rabbet

There are literally thousands of different joints, most all being variations of the aforementioned three. Whatever type of joint you elect to use, it has to be well prepared first. The pieces will have to fit properly. Gluing will not correct a lot of mistakes; it will only hold two pieces of wood together if they fit properly in the first place. The wood surfaces have to be smooth, clean, free of dust, old glue, oil, paint, varnish or other types of finishes or anything that could interfere with the adhesive.

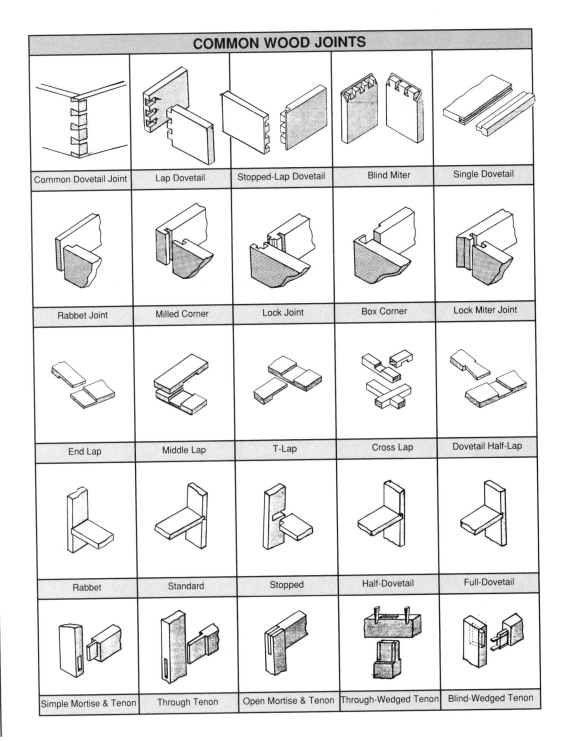

COMMON WOOD JOINTS

Common Dovetail Joint	Lap Dovetail	Stopped-Lap Dovetail	Blind Miter	Single Dovetail
Rabbet Joint	Milled Corner	Lock Joint	Box Corner	Lock Miter Joint
End Lap	Middle Lap	T-Lap	Cross Lap	Dovetail Half-Lap
Rabbet	Standard	Stopped	Half-Dovetail	Full-Dovetail
Simple Mortise & Tenon	Through Tenon	Open Mortise & Tenon	Through-Wedged Tenon	Blind-Wedged Tenon

DOWELS & DOWELING

The use of dowels is still very popular. Dowels are used to align and reinforce a glued joint. When they are properly placed, they help to line up the pieces of stock and prevent the pieces from drifting or shifting out of place. They are often used as a substitute for a "mortise-and-tenon" joint, a "spline" joint or a "tongue and groove" joint. A dowel joint is not as strong as a mortise and tenon joint, but it is easier to prepare.

Common dowels are usually round pieces of Maple or Birch. Dowels made from Oak, Walnut and Mahogany are also available. Dowel rods are available in 36" lengths and range from 1/8" to 1"+ in diameter, increasing in diameter by increments of 1/16". It is most important, if you purchase a dowel rod to make your own dowel pins, that you cut grooves into the side of the dowel pin by running the dowel pin over a saw blade, or forcing it through a dowel former. The dowel former sizes the dowel to an exact diameter while at the same time cutting small flutes or grooves in the face of the dowel.

Dowel pins are also available. They are pre-cut lengths of dowel rod having a straight or spiral groove cut into the outside face of the dowel.

These grooves are extremely important as they allow any excess air and glue that is compressed in the hole during assembly to escape. Without such grooves the principle of hydraulics could cause the air and glue trapped in the dowel hole to be compressed, causing the face of the wooden pieces to explode or burst due to the extreme amount of pressure exerted during final assembly and clamping.

The ends of the dowel pins are rounded over or chamfered to facilitate inserting the dowel into its hole during the assembly of the wooden pieces. It also helps to prevent the ends of the dowel from splitting or being burred over when being struck with a hammer or mallet. This contouring of the dowel ends can be quickly and easily accomplished on small diameter dowels using a pencil sharpener or dowel former. A plane, file or sandpaper may also be use to create this chamfer.

When making a dowel joint, select a dowel with a diameter equal to half the thickness of the smallest member of the joint, i.e. use a 3/8" dowel for 3/4" wood or a 1/4" dowel for 1/2" wood.

The length of the dowel should be such that the dowel will enter the hole in each piece a depth equal to 2-1/2 times the diameter of the dowel.

Dowel rods range from 1/8" to 1"+ in diameter.

If you cut your own dowel pins, use a dowel former to cut grooves into the pin's surface.

Dowel pins are pre-cut to the most frequently used lengths.

The spiral or straight grooves allow excess air and glue to escape. The chamfered ends facilitate insertion of the dowel into its hole.

When joining two boards together on the edge, the dowels should be placed approximately 8" to 12" apart, and no dowel should be any closer than 2" from the end of the board. When joining leg and rail assemblies for chairs or the leg and skirt assemblies for tables, the dowels may be as close as 3/4" on centers. The final decision for the placement and size of dowels ultimately depends upon your own good judgement.

Patented doweling jigs are available to assist you in positioning and drilling the holes for your dowels.

A large capacity doweling jig.

Turret-style doweling jig.

You may consider using a doweling jig if you are going into production work, or you may set up a drill press and accomplish the same results. A "Portalign" accessory tool is available to attach to your hand held electric drill. This will guide your drill bit perpendicularly into the wood, or it may be set for any desired angle. Be sure to read and follow the manufacturer's instructions for any of these jigs.

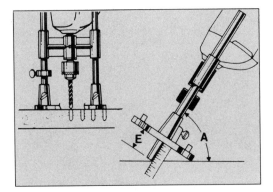

Portalign Drill Guide

50

For the woodworker who only uses dowels occasionally, consideration should be given to using "dowel centers" to assist you in aligning opposite holes for your doweling project. Dowel centers are round pieces of metal, available in various diameters to match the diameter of the dowel that you have selected for use. One end of the dowel center has a small shoulder faced off with a small pointed protrusion. These dowel centers are available in sets of various diameters, or you may select them individually by specified size.

Select a dowel center with the same diameter as your dowel rod or pin.

Select a drill the same size as the diameter of the dowel you are going to use. If boring with a hand brace, use an auger bit; if using an electric drill or a drill press, use a brad point spiral drill or a spade type drill bit. Do **not** use a drill bit made for drilling metal.

NOTE: DO NOT USE "ALL-PURPOSE" DRILL BITS DESIGNED FOR BORING METAL.

Auger Bit   Spade Bit   All-Purpose or Metal Bit
Brad Point Bit   Forstner Bit

Mark a pencil reference line across the two members of the joint.

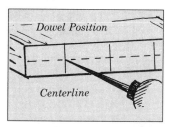

Dowel Position

Centerline

Mark out the dowel positions on the edge of one of the pieces. Use an awl to locate the center of your dowel hole.

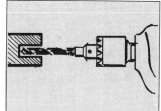

Drill a hole with a diameter equal to that of the selected dowel and a depth equal to half the length of the dowel plus 1/8".

Carefully align the two surfaces that you are going to join. With a pencil, mark a line across the joint of the pieces that you are going to join, this will be your reference line. Mark out on the edge of one of the pieces where you want your dowels. Use an awl and indent the center of the dowel position you have selected.

Drill a hole to a depth equal to 1/2 the length of the dowel you are going to use, and add 1/8" for bottom clearance. Be sure to hold the drill perpendicular or vertical to surface you are drilling (this is where the use of a doweling jig or the Portalign comes in handy), unless a specific angled hole is needed.

After all holes are drilled in one of the pieces, insert the selected size dowel centers into the holes, and realign the two surfaces of the joint, checking to see that your reference line is in alignment. Press or tap the two surfaces together, resulting in a small depressions marked in the mating pieces. These depressions are now the centers for your matching holes, which you can now drill.

Insert a dowel center into each dowel hole and use your reference line to realign the two members of the joint. Press or tap the two pieces together.

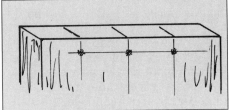

The tiny depressions left by the dowel centers mark the locations for your matching holes.

After all holes are drilled, slightly countersink the edges of the holes (no more than 1/16" deep). This will remove any burrs caused by the drilling and also leave a small clearance for your glue. Insert the dowel pins into holes without using any glue. Press the two pieces of the joint together and you should have a perfect joint. Disassemble all pieces in your assembly and prepare to glue up for the final joining of the pieces.

Use a small stick or brush and apply a **thin** coat of glue to the insides of the holes on one piece of your assembly. Apply a thin coat of glue to the ends of the dowels and, with a mallet, drive the dowels into the holes. Spread a thin coat of glue on the two surfaces to be joined; also apply a thin coating of glue into the exposed holes and dowel ends.

Proper gluing method

Align the dowels and corresponding dowel holes, then press the two pieces together. Fasten the joint tight, using clamps. Check with a square or straightedge to be sure the members of the joint are in alignment. Wipe off excess squeezed out glue with a clean cloth dampened with water. Place the clamped assembly aside to allow the glue to cure and dry.

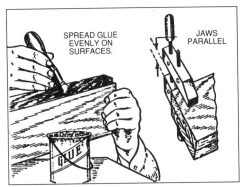

TIPS & TECHNIQUES

1 Dowels are wood and, while in storage, absorb moisture, swell in diameter and are very often oversize. Size all dowels with a dowel former. Don't try to force an oversize dowel into a hole; it will split the wood.

2 Be sure the dowel pins are chamfered, grooved or striated.

3 Be sure the dowels are shorter, approximately 1/4" less than the combined overall depth of the two matching holes.

4 Unless otherwise required (angle drilling), all drilling should be done perpendicular or vertical to the surface of the hole. You can use a square resting next to your bit while drilling to check that you are drilling properly.

5 Use depth collars on your bit to be sure that you drill the proper depth. A piece of masking tape wrapped around the drill bit at the proper depth can be used to indicate depth of hole while drilling.

6 When checking your assembly dry, without glue, the dowels should fit snug, not too tight or too loose.

7 Do not depend upon an excessive amount of glue to secure the bond. A thin layer of glue on a properly prepared joint is all that is necessary.

8 Don't depend on clamps to close a badly fitted joint. Bad joints can be closed with excessive clamp pressure, but large gaps will eventually open up again.

STRUCTURAL REPAIRS

The following paragraphs will describe in detail some of the techniques in repairing furniture. Throughout the text, reference is made to chairs; however the techniques, instruction and methods are applicable to most any piece of furniture.

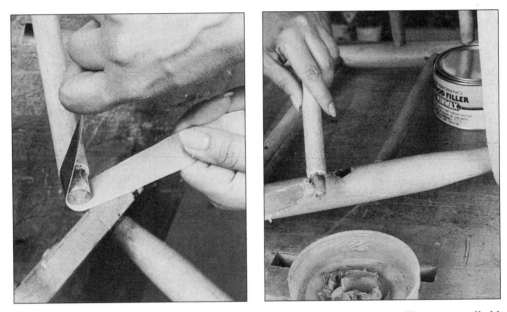

Regluing rungs, legs and stretchers is a common and relatively easy repair. First remove all old glue. If rung is loose, wood filler may be used.

Unsafe and unsightly chairs are in abundance. Some are found around the average home and many in secondhand furniture shops. Quite often, you can pick up real bargains by talking your way into the back rooms and storage barns of antique shops where the owner has relegated chairs needing replacement parts, a task he has postponed, and he is now prepared to accept a token price for the item. Chairs usually are the best bargains and the hardest pieces to restore. But antique chairs of good design are well worth your trouble. Here are some guidelines for chair repair.

The back legs and the joints on the back legs of a chair are subjected to great strain, and go first, especially when you sit on the chair, relax and tip back on the back legs (a No! No!). If the break is a bad one, especially in a rear leg, it may be necessary to make a new leg, although in the case of antiques of real value, this should be avoided if possible. It entails a lot of work because the old leg must be taken off by methods described in earlier discussions (see disassembly). Sometimes the repair can be accomplished on one section of the leg without dismantling the chair. This may be done by cutting above or below each joint and splitting the wood away from the tenon or dowels. If the repair is sectional or a total replacement is necessary, the opposite good leg may be used in preparing a pattern for making the new part.

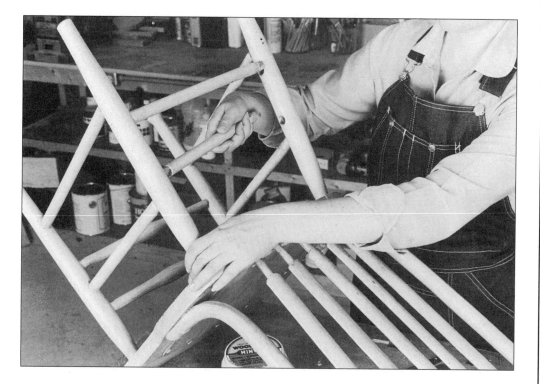

Quite often it is just one part of the defective leg that needs to be replaced. Various types of joints are suitable. Your choice depends on the leg structure. Several options are illustrated.

Dowels may be inserted to align and reinforce the joint. Experts prefer to use two small dowels rather than one large dowel. Of course all such joints are assembled with quality furniture type glue, not the white type craft glue.

When splicing a section to a leg it is good practice to make the replacement section somewhat larger in cross section than required. This allows for final trimming, carving and shaping if necessary, after the pieces are assembled, so that the new part will match and blend into the old section.

It will be necessary to replace the entire member if the pieces are missing or in such bad condition as to forbid splicing of a new section. Turnings can be made on a lathe, but there are situations when plain spindles, either tapered or straight, can best be shaped by hand. If it is handwork, the part can be worked from a square piece of stock using a block plane, rasp and sandpaper.

Observe that all these repairs are variations of the spliced joint. The spliced joint, by its very nature of overlapping new and old sections, is by far the strongest for this purpose.

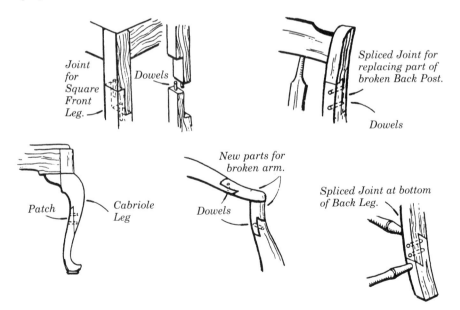

Joint for Square Front Leg.

Dowels

Spliced Joint for replacing part of broken Back Post.

Dowels

Patch

Cabriole Leg

New parts for broken arm.

Dowels

Spliced Joint at bottom of Back Leg.

REPLACING BROKEN OR MISSING RUNGS, SPINDLES OR SPREADERS
Without Disassembling Sound Joints.

Clean all waste and old glue from both holes. Rebore only one hole somewhat deeper, taking care not to drill through the piece. A flat bottom drill such as a Forstner bit is good for this.

Prepare the replacement piece so that one end will slide deep enough into the rebored hole to allow the other end to just slip into its matching hole. Slide the end of the replacement piece to the bottom of this original hole and you have replaced the piece without damage to any other joints. After checking assembly for fit, etc. disassemble and use your selected glue and reassemble the piece for the final time.

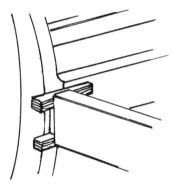

Hold joint open with wedges.

Very old chairs are often victims of crystallized glue and wood shrinkage. Loose legs, the most common problem, can sometimes be corrected without complete dismantling, especially if the repair work is taken in time. The joint between the rail and leg is usually the first to weaken. If one joint is quite loose but others are tight, try to force glue into the loose joint.

Drive two thin wedges to open the joint slightly. Take great pains not to strain the good joints that are still tight. When the tenon or dowels are somewhat exposed, carefully scrape away the old, crystallized glue as much as possible without damage to the wood or the surrounding finish. Spread on fresh glue with a sliver of veneer. Use a resorcinol-resin glue even though it is a little more trouble to mix glue than to use a "ready mix" glue. Elmers Waterproof glue is one brand to be considered. This glue has marked gap-filling ability, ideal for loose chair joints. Apply suitable clamps protecting wood surfaces, wipe off squeeze out and allow assembly to dry. Titebond glue has also been used very successfully to tighten loose chair joints.

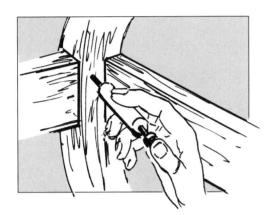

If a joint is wobbly but not loose enough for the wedge technique, insert glue by means of a glue injector. Drill two small holes into a concealed inside surface of the leg. Make a guess that the joint is held with two dowels and aim for each dowel. Inject glue into these holes, wipe off squeeze out and apply clamps. The holes may be patched with stick shellac or wood filler during the finishing schedule.

After either of the above repairs has been made, the next best aid is a corner glue block. The joints between rear legs and rails come under the greatest imaginable strain, especially the legs of antique chairs which often were built without lower stretchers. If the chair has a removable slip seat, remove the seat to provide working space for installation of a glue block.

Many chairs already have a corner block. Even then, rear chair legs can work loose in time. Wood shrinkage and old glue that has crystallized are common causes of joint failure. Probably the glue block is no longer doing its work. It can be made to work better. The first step in the repair is the removal of the block. With a little change in these blocks, they can be made to hold the leg in place far better than before.

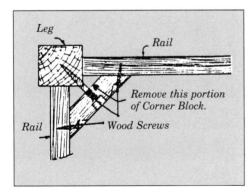

Cut off a portion of the block along its shortest face so that when replaced it will not touch the leg but will still pull the rails tight. Locate a wood screw long enough to pass through the block into the corner of the leg. Bore a hole in the block for passage of the screw. Spread glue on the corner block where it contacts the rails. Ready-to-use yellow glue is satisfactory. Drive the long screw to pull the leg up tight. But before tightening the screw fully, flow a little glue into any gap that may have appeared between rail and leg. Use a glue injector, thin knife blade or scrap of veneer to push glue down into crack.

This procedure, of course, is not adequate for a leg that can be pulled off when glue block is removed. In that case complete dismantling is required (see below).

A fairly recent innovation which replaces the "glue block" is the Hanger Bolt and Corner Brace. A proper size hole is drilled on the inside corner of the leg and a hanger bolt is screwed into this hole. Next saw kerfs are made or small thin blocks of wood are fastened to the inside of each rail; the kerf or block is properly positioned on each rail depending upon the size of Corner Brace used. After the wood joint (dowels, mortise & tenon, etc) is glued and assembled, the brace is placed over the hanger bolt with the edges in the kerf or over the edge of the small block, the nut is fastened on the hanger bolt, and the assembly is tightened. You have just fastened and married the three pieces of the assembly.

Hanger bolt and corner brace assembly.

When a chair leg is wobbly but not loose enough to come off, yet too loose for for satisfactory repair by previous discussed methods, dismantling is called for.

To remove a wobbly leg, work it gently in all directions. If this won't free the leg the next method is to use water (see "disassembly"). Of course the water treatment may cause some damage to the finish and require some finishing touchup. However, if the chair is old, it is safe to guess that the original glue is not water resistant. The water solution will soften the glue enough for your purpose. The use of live steam is another technique used to soften old glue. Run a rubber hose from a boiling tea kettle to apply steam to the joint. This method is effective but hazardous and often awkward.

Sometimes it is essential to cut a stubborn joint. A repairman ahead of you may have tightened the leg structure with a hidden wedge driven into the end of a tenon. In this case there is no alternative to cutting the joint. The trick here, is to work a very thin saw blade,

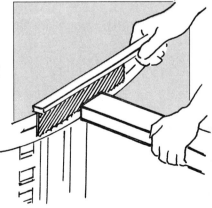

dovetail saw or coping saw, into the joint without scuffing the adjacent areas. Insert the saw into the wide face of the tenon, not into its narrowest thickness. This action will often split the tenon in the grain direction and separate it in half to permit easy withdrawal.

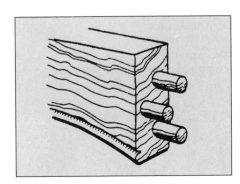

The remaining stub can be carefully drilled and chiseled out of the mortise or socket. In this case the socket is then plugged with new wood, while the tenon is cut off flush so the joint can be doweled. The sawing method can be used also to cut through a doweled joint after all other methods fail. Remaining dowel stubs may then be cut flush and new holes bored for new dowels.

When a tenoned rail is loosened sufficiently to be pulled out of a socket in the leg without resorting to the saw method, the tenon probably is too loose for simple gluing and reassembly. The wood may have shrunk and the tenon now should be built up. If the joint fits loosely, other methods are advisable. To tighten a mortise and tenon, for example, the tenon can be built up to fit tightly. Veneer serves well for this purpose. Avoid veneers that have a tendency to split. Locate a scrap of mahogany, birch, cherry, poplar or a wood of similar characteristics. Veneer should be glued to each of four surfaces of the tenon. Adding it to only one or two surfaces, especially adjacent faces, will move the entire rail off center. When the glue is dry, trim down the enlarged tenon to fit the mortise.

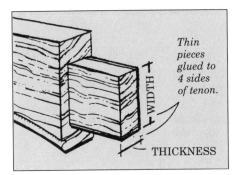

Thin pieces glued to 4 sides of tenon.

WIDTH

THICKNESS

If the tenon is damaged and cannot be restored with reasonable measures, it should be cut off entirely and dowels may be used for reassembly of rail to leg. Of course the mortise must be plugged with a tight fitting block which is to be planed flush and bored for dowels.

Many of the old "Kitchen Hitchcock" chairs are still around. Legs usually had rounded tenons fitted in blind holes in the underside of the seat. When these legs become loose, remove them and use the wedge method of

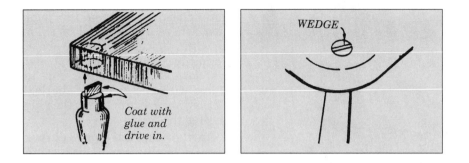

Coat with glue and drive in.

WEDGE

tightening. If originally wedged,the wood has shrunk and now needs a thicker wedge. Cut a narrow slot in the leg to receive a thin wood wedge, the length of which equals the depth of the slot. Push the wedge only partially into the slot. Force this assembly back into the hole. Drive it tight with a rubber mallet. (Of course the old glue was removed prior to the application of your new adhesive.)

Some of the old country furniture had a leg tenon coming through the the seat. If now loose, remove the wedge. A narrow straight chisel used carefully will remove enough of the wedge for withdrawal of the leg. Reassemble with a somewhat thicker wedge.

TIGHTENING LOOSE JOINTS

It is not surprising that chair and table joints break down even when they are not abused. These joints are subjected to great strain. When a table or chair develops loose joints, adding more glue is not always the best way to make the repair. Regluing may be satisfactory if the problem arose because the old hide glue crystallized.

But regluing will only work if the joint is still fairly tight. Of course the joint must be taken apart and scraped clean with a knife or chisel. (The use of sandpaper to remove old glue is **not** recommended. This leaves wood dust and grit in the wood pores.) Simply clean, reglue and clamp.

A loose doweled joint ordinarily requires replacement of dowels. Bore them out if they are still tight in one part of the joint. Usually the rail and leg are large enough to allow for boring larger holes and inserting the next size of dowel.

Plugging one of the oversize holes is another remedy. This method is often used to tighten the joint formed where a round rail or chair rung with a doweled end is involved. Fit a hardwood plug in the hole and dress down the projecting end to conform with the shape of the leg. Drill the proper size hole into the plug to receive the round rung.

Another suggested procedure is to wrap a small piece of cheesecloth (do NOT use synthetic materials) around a lightly glue coated dowel or tenon of the joint. Use just enough cheesecloth for the dowel to fit snugly into the hole. Cover cheesecloth with a light coat of Titebond glue and force back into hole, trimming off any excess cheesecloth if necessary prior to reassembling joint. Since cheesecloth is cellulose and so is wood, you are combining the same materials.

BORING OUT OLD DOWELS

If new dowels of the same size as the old ones are to be used, it is best to bore out the old dowels with a bit one size smaller than the existing original hole. Next, use a sharp-pointed tool such as an awl to pry and chip away the remaining wall of the old dowel, breaking the pieces towards the inside of the newly drilled smaller hole. A small gouge may also be useful in cleaning the hole. In this way, the original hole can be used again. On the other hand, if an attempt is made to bore out the old dowel with a bit of the same size, the bit is likely to run off center or at a slight angle and thus cause difficulty in assembling the work.

WOOD SELECTION

When making new parts for antique chairs it is necessary, of course, to locate wood that will match the old wood in color, figure and texture. With genuine antiques of real value the best practice is to repair parts, if possible, without inserting new sections or whole parts. When replacement is essential, the ideal source is another badly damaged chair which can be your supply for the same kind of wood. Occasionally a repair shop will have a suitable piece of discarded wood available at a bargain price.

Usually, however, you will have to procure a piece of new wood of the same kind. Then, using modern finishing materials and methods, you can blend the new wood into an amazingly skillful match with the rest of the old wood. For the best results, try out your finishing schedule on a piece or two of scraps cut from the stock you are using for the repair.

CLAMPING

Clamping chair parts together often involves a lot of experiment because the surfaces are neither straight nor parallel. Judicious placement of clamps (versatile wooden-jaw hand screws, the cabinet makers favorite), will usually provide a solution for your clamping problems. Two of them may be fastened on either side of a joint to give a gripping surface for other hand screws, or a bar clamp may be used in actually drawing the joint together.

Bar Clamps and hand screws in use.

At this point in restoration work, with the chair in clamps, you might do well to follow a trick some knowledgeable repairmen use to avoid the exasperating "short leg." They place the chair on a perfectly level floor or plywood slab and lay a very heavy weight, such as bag or bucket of sand, on the seat. The weight makes certain the chair will rest four-square on the floor when the glue hardens.

Band Clamp

BENDING WOOD

Replacing parts that are bent to shape requires a lot of work and ingenuity. The new part must be made from straight stock, then bent by steaming the wood and inserting the steamed piece into a fixture that firmly holds the wood to the curved shape until it is dry.

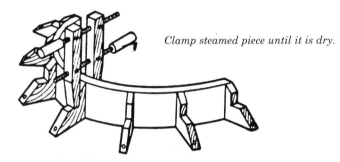

Clamp steamed piece until it is dry.

Steaming and bending of wood is a specialized subject by itself, often requiring large bulky equipment. For bending small, thin pieces of wood, the steam jet from a kettle may be sufficient. Another method for bending small pieces requires a metal container large enough to hold the new square of wood. The container is set on a stove, filled with water which is heated until it boils. The wood must remain in boiling water for several hours, the time depending upon the size and species of the piece and the extent of the bend. It may be necessary to add water from time to time. Have the appropriate bending form ready to receive the soaked wood. Leave the piece in the form several days or longer to dry thoroughly. Old Windsor chairs nearly always need replacement spindles and side posts that require bending, but a beautifully restored Windsor is an enduring reward for your efforts. Exercise great care as live steam can cause severe burns to your skin.

SURFACE REPAIRS

FLATTENING VENEERS

Whether you are making repairs to a veneered surface, or veneering the entire surface, the veneer that is to be applied must be flat. Due to the nature of the veneer manufacturing process, the veneer, especially Burl veneers, is often supplied to the consumer buckled or wavy. Only experience will tell whether a buckled veneer can be laid satisfactorily without preliminary flattening.

The following method may be used to flatten warped veneers.

Prepare a sizing solution of: 1/4 cup Hide Glue Flakes, +1 gallon of Water, + 6 oz. of Glycerine, + 6 oz. of Alcohol. Put the solution in an empty spray type bottle and moderately spray both sides of the buckled veneer. Place the veneer between two sheets of brown wrapping paper (not newsprint). Lay the assembly on a flat surface and cover with a weighted flat panel, or use a veneer press. Examine the following day. If the veneer is not flat, spray again and repeat until the veneer is flat. When flat, do not spray any more solution but continue to press it for five more days, changing to new, dry paper daily to withdraw all the moisture that has been added. Keep the veneer flat and covered at all times when not working on it.

The glue will strengthen and size the veneer, the glycerine will flexisize the veneer, the alcohol will help evaporate the water and the water is a vehicle for all these materials. If the paper sticks to veneer, gently scrape or sand it off, taking care not to remove any of the wood veneer.

Veneer needing flattening.

Apply water with sprayer.

Place between sheets of paper (towel or brown).

Layers of paper-wrapped veneer.

Weight with boards.

Pail of sand for added weight.

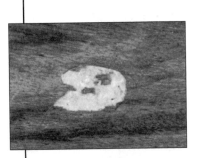

Damaged Veneer Surface.

Square off all edges and contour surface of remaining intact veneer.

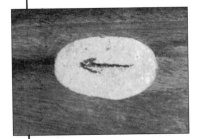

Mark and Index position of patch in recessed area.

BROKEN OR MISSING VENEER

One of the most common forms of damage to a veneered surface is a broken off or a missing piece. Usually such damage is found on the edge of a veneered surface; however, such damage can occur within the veneer surface itself.

When a piece of veneer is broken, it does not break sharp and square, but leaves a rather tapered edge from the bottom to the top surface. First it is necessary to square off the tapered and rough edges of the old veneer that remains bonded to the surface. The edges of this remaining veneer should be perpendicular to the surface to which the veneer is bonded.

Using a craft knife with a pointed blade, carefully cut the remaining ragged edges away from the damaged area so that you have a clean cut hole in the veneer surface and the edges of the remaining veneer are perpendicular or square to the bonding surface.

When cutting across the grain of the veneer, do not make a sharp straight cut, which will be obvious when the patch is complete. Rather make a sweeping circular cut across the grain; this circular cut will be less obvious.

It is essential to remove all traces of any old adhesive on the surface upon which the new veneer patch will be bonded. New glue does not hold to old glue. Careful scraping with a knife, hand router plane, or chisel will accomplish this.

Next be sure that the surface to which you are going to bond your patch is perfectly level. If necessary, use a wood filler such as Elmers Carpenters Wood Filler to fill in any gouges or voids in the bonding surface. Be sure filler is completely dry before proceeding. Any imperfections in this bonding surface can show through to the top surface of your new veneer patch.

You should now have the edges of the remaining veneer square or perpendicular to the bonding surface and this bonding surface is now perfectly level and free from all old adhesive and dirt.

You are now ready to make an exact pattern of this area that you have prepared. Tape a piece of plain white paper over the area and, using the side of a soft pencil, make a rubbing of the area. This will give you an exact pattern of the shape of the recessed area. With your pencil, mark an arrow in the center of your pattern, remove the tape and pattern, and mark a similarly oriented arrow in the recessed area of your veneer surface. This will make it easier to properly place the patch into the recessed area, especially if the shape is symmetrical.

Now it is necessary to select a new piece of veneer that will closely match the remaining veneered surface. Select a piece of veneer of the same type of wood, same thickness, similar in color, grain and texture.

Apply a coat of rubber cement to the back side of your pattern or rubbing that you have made. Fasten this pattern firmly over the selected area of your replacement veneer using several passes of your veneer roller. With your craft knife, lightly cut along the outline created on your rubbing. Favor cutting to the outside of the line. You can always cut away any excess material if necessary. Many light passes of the knife will give you a cleaner edge than trying to cut through the veneer in one or two passes. Do all your cutting on a scrap piece of wood or on stiff cardboard so that when the point of the knife does penetrate the veneer, it will not break off the point.

When the patch piece is completely cut out, check the patch for fit in the recessed area you have prepared. If necessary carefully trim your patch with your craft knife or a fingernail emery board. Next, carefully peel back and remove the pattern from the patch piece. By using rubber cement to bond the pattern to the veneer, it will easily peel off.

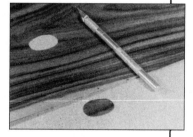

Cut patch from wood veneer similar in color, grain and texture.

Your patch should fit perfectly. The grain of the patch should line up with the grain of the surrounding area. If the patch does not fit snugly or you are not satisfied with the match of the grains, make a new pattern and cut a new patch. Once you are satisfied with the patch, you are ready to bond the patch into place permanently.

Use either a white or yellow glue. ***Do NOT*** use a contact cement. Place a light coat of the glue into the recessed area, being sure to get the adhesive on the edges of the remaining bonded veneer. Place a very light coat of the glue on the back side of your patch. Place the patch into the recessed area; using your veneer roller, roll the patch firmly. With a clean cloth dampened, not soaking, with water, wipe off the excess or squeezed out glue, place a piece of waxed paper or plastic over the area, add about eight sheets of newspaper that will act as a cushion, then apply a block of wood on top of this assembly and clamp or add weights and leave overnight.

Install patch into recessed area and roller flat.

The following day, remove clamps or weights, newspaper, waxed paper or plastic and you can then examine the results of your efforts. Proceed to properly sand the area, stain if necessary and add your finish. If you exercised care in cutting and selection of the veneer for the patch, you should have an almost invisible patch. Remember, you will always know where the patch is, but the crucial test is, can anybody else find it?

REPAIRING BLISTERED VENEER

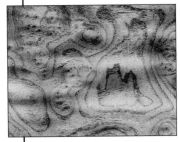

To repair blistered veneer...

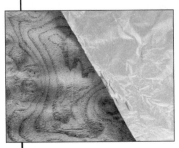

Place foil on blistered area...

Then apply heat with an iron.

When a blister occurs in a newly laid veneer, there is a fairly good chance of correcting the damage in a simple way. This remedy is appropriate for blisters and larger raised areas where the veneer is not broken. If you constructed the furniture concerned, even if construction took place a year or two ago, the remedy suggested here is well worth trying. However, if the problem shows up in an old piece, most likely the glue beneath the blister has crystallized and can not be reactivated. Another technique must be followed.

Here is the simplest repair and it is worth trying first. If sufficient glue was originally spread on the core during construction, and if the glue is not yet old enough to have crystallized or been subjected to excessive heat, the existing glue can probably be reactivated by the application of heat.

Borrow the household electric iron. You will need a square of aluminium foil from the kitchen and either a wooden veneer roller or a veneer hammer. Have on hand for immediate use a caul board, such as a square of particle board only slightly larger than the area being treated.

Figure out in advance a means for clamping the spot you are repairing. On the top surface of an existing table, clamping can be a difficult problem. If not solvable, collect some heavy weights to use instead of clamps. Weights are not as effective as clamps, but you have no alternative. On curved or irregular sufaces, fill a heavy duty plastic bag with a quantity of sand and place this on the irregular surface to provided the necessary weight. The bagged sand will conform to whatever surface it is placed upon. Also have ready a square of brown paper bag without print or seams.

Set the pressing iron for moderate heat. Lay aluminum foil over the blister. Rub the iron back and forth. **Caution! DO NOT let iron remain stationary on the veneer surface.** Work over the blister and not much beyond it. Lift the foil away, replace it with the square of brown paper, and immediately apply pressure with a roller or veneer hammer. The brown paper will protect the wood and finish while you apply heavy pressure. Test the blister and repeat the procedure even though the blister seems to be sticking down. Then lay the brown paper in place again, put the caul board or wood block on top, and set up your clamping system or weighting arrangement. Leave the setup at least overnight. A hair dryer may be substituted as a source for heating the area to be repaired.

If sufficient glue was originally spread on the core, the cause of the blister was probably too little clamping pressure at the blistered spot at the time of construction. Glue that has not yet crystallized will respond to heat and pressure applied as described. The preceding remedy should lay down the blistered area. If the blister still is visible other tactics are called for.

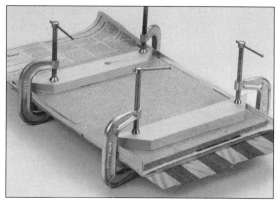

A blister that did not respond properly to the heat treatment will now require glue. You will have to slit the veneer to introduce glue under the blister. It may seem proper to slit the veneer at the middle, but to do so

Clamp the repair overnight.

and then inject glue will cause the glue to mound mostly under the slit and not spread to the edges of the blister. The right way is to make razor slits at opposite sides of the blister. Only a large blistered area will require a slit also in the middle.

If the side slits permit, lift the veneer very gently so you can work a very thin knife blade, steel spatula, pin, needle, or dental pick underneath to loosen crystallized glue. If the furniture is old, most probably hide glue was used. This can be softened by squirting a few drops of warm water and white vinegar through the slits (a glue injector is good for this.) Allow the water and vinegar solution to dry up and disappear. Suspend an electric bulb above the area as a mild heat source to hasten drying. If the slit is long enough perhaps you can work a spatula wrapped with a layer of lintless cloth into the interior for drying and cleaning. However, it is better to let time dry the area rather than risk a split in the veneer.

Next, use a glue injector and a small spatula to work new glue underneath the blister. Yellow glue is recommended for this repair. Use glue sparingly to avoid excessive squeeze out. The important job here is to get the glue evenly and thoroughly spread as far as you can safely work. Wait two or three minutes for the glue to tack before

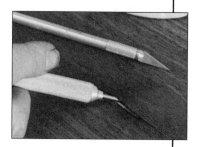

Slit edges of blistered area and inject glue underneath.

you apply finger pressure gently over the blister. The veneer will absorb some moisture from the glue and should press down without splitting. When the blister eventually levels, lay brown paper on top and roller veneer into place, first gently and finally firmly. When all danger of splitting has passed, wipe off squeeze out, and immediately set up your caul board, newspaper cushion and clamps or weighting system. Leave clamped or weighted assembly at least overnight.

The above method for blister repair should work very well. In the unlikely event of further trouble at the same spot, due most likely to inadequate rollering or too little weighting pressure, you could resort to the heat treatment described earlier. Now that there is some new glue under the blister the heat treatment has a good chance of solving the problem. In old furniture the cause of blisters is glue crystallization either from age or from standing too near a home heating source. In newer furniture the causes are usually too little glue or too little pressure when laying down the veneer. Some of these causes can be avoided by proper initial application of the veneer.

REPAIRING BUCKLED VENEER

Large raised areas of veneer, considerably more buckled than the small blister, will not respond to the electric iron trick. If this type of damage is too extensive for slitting and working glue underneath, a third repair method should be tried. The badly buckled veneer can likely be flattened. If the first method does not work, the buckled area will have to be cut out and either flattened before being repaired or completely replaced with new veneer.

To flatten buckled veneer that has lifted from a surface to which it had been glued, place a clean, damp warm rag over the buckle. (Flattening a square or sheet of veneer that has not been laid down calls for another treatment — see "Flattening Veneer." After a few minutes of moisture penetration, lay a caul board such as a square of particle board over the rag, with waxed paper under the caul to prevent it from getting damp. After a few more minutes, replace the damp rag with a warmer one but not a wetter one. Start now to add moderate

To flatten veneer, you will need some damp rags, paper, a caul board and weights.

weight on top of the caul. Keep this system going until the veneer becomes flat enough to add considerable weight without danger of cracking. Leave assembly overnight. Repeat the procedure the next day if the veneer is not yet straightened. When the veneer has become fairly straight, cover it first with clean wrapping paper, followed with a cushion of dry newspaper, a caul board and heavy weights. Replace the wrapping paper and the newspaper frequently as they become damp.

When the buckled area has become dry and straightened enough to be pressed flat, slit it to create enough opening to scrape loose some of the crystallized old glue. Use any practicable easy way to remove the glue dust, including a vacuum cleaner. When the area is clean, work new glue under the raised veneer using a glue injector and a small spatula. Yellow glue is recommended. Use fresh glue because it is thinner and more manageable for a difficult job than is bottled glue that has aged a couple of years. Quite possibly slits will have to be made at edges of the buckle and a slit in the center.

Allow at least three to five minutes for the glue to tack so as to reduce the glue squeeze out to a minimum. Remove squeeze out with the point of a knife to avoid smearing. Finally use a damp cloth with utmost discretion to remove excess squeeze out.

Sometimes in severe buckling a more drastic type of slit may be necessary to assure complete glue coverage beneath the raised veneer. Use a two sided slit similar to the two sides of a triangle. This technique gives access to more gluing area. To minimize danger of splitting, these major cuts can be made while the veneer is still damp and pliable, before you start the drying procedure. Follow gluing, rollering and clamping or weighting methods described for difficult blisters.

After blisters and buckles have been repaired, the knife slits made for glue can be filled if they show up too much. Make a mound of matching sawdust from scrap of same or similar veneer, mix with **white glue** and pack this filler firmly into the cracks. Carefully wipe away surplus squeeze-out, allow overnight drying, carefully sand and then finish.

REMOVING DENTS IN RAW WOOD

Unfinished solid wood may be found to have a dented damaged spot. As long as it is only a dent (compressed wood fibers) and not a gouge, cut or scratch (wood fibers severed or removed), there are several ways to salvage the piece. For severe deep dents, various types of wood fillers or shellac sticks may have to be considered, but the simplest of all remedies should be tried first. Water!

It is well known that water and steam will make wood swell. By applying this principal to the dent in raw wood, you may be able to salvage a valuable piece of wood. The one point to keep in mind is restraint. First, with a sharp pin or pointed knife, put several small holes in the bottom of the dent. Apply a few drops of water in the dented spot only, allowing it to soak into the wood. An eye-dropper or glue injector will serve the purpose best. If the mark is not too deep, a single application will be enough to bring the surface back to normal. On the other hand, should the indentation be quite deep it may be necessary to apply water several times, allowing at least a half hour between applications. Sanding the area the next day is usually sufficient to restore the damage. The trick is worth trying on a piece of wood you want to use and of course, on a piece of unfinished furniture you have constructed.

To raise a dent, first pierce the wood with a pin then apply a few drops of water.

A steam iron may also be used. Place the steam vents over the dent (which you have pin picked at the lowest portion of the dent) and release the steam for a few seconds. Allow area to dry and repeat if necessary. Take care that the iron is not too hot as it may scorch the wood surface.

Another trick—apply a clean wet rag over the damaged area. Press a hot soldering iron on the wet rag over the damaged area. This will create steam which, in turn, will expand the cells of the compressed wood to their original shape. Allow to dry thoroughly overnite before sanding and further finishing.

FILLING CRACKS and HOLES

To fill a small hole, make a pile of scrapings from the wood.

Mix with white glue and press into hole.

For filling small holes and cracks on wood surfaces that are not going to be stained, try the following procedure. With a single edged razor blade, scrape and make a small pile of scrapings of the wood you are going to fill (take from the underside of the piece, if available). Mix these scrapings with **White Glue** to the consistency of heavy cream. Press this mixture into the hole or crack. Level off to slightly above the surrounding surface; allow to dry overnight. White glue dries transparent; what you have visibly left is the color of the wood scrapings and you can now proceed to sand and finish the entire surface.

SHELLAC STICKS The art of using shellac sticks is one of the most useful skills in furniture repair. The skill is not acquired easily and should be practiced on some unimportant piece of furniture or on scraps of wood. A repair made with shellac sticks can look blotchy or it can be virtually invisible. There are colors to match practically any furniture finish. When two shades of shellac stick are close to the finish you are matching, choose the lighter shade.

Shellac sticks come in a variety of colors to match almost any wood.

Shellac quickly becomes molten when touched by a hot steel spatula, but it also hardens very fast. A small, flexible, steel spatula, often referred to as a burning in knife, makes the best applicator. The end of the knife is heated in a carbonless flame, an alcohol lamp. The heated end of the knife is quickly touched to the shellac stick and a small quantity of the shellac is held on the knife and then transferred into the bottom of the damaged area of the wood surface. Do not attempt to fill up the void in one

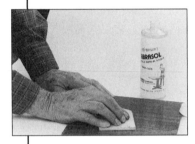

application of the shellac. Make several thin applications.

As the layers are applied, they will melt together. Use the heated knife to press the melted shellac into the damaged area. Apply enough shellac so it is just slightly above the surrounding surface. Use the heated knife to level this final, slightly crested surface. It is obvious that this operation can cause damage to the immediate surrounding area. This area can be protected prior to starting the repair procedure, by applying a small quantity of "Burn In Balm" around the damage. An Electric "Burn-In-Knife" may be substituted for the spatula and the alcohol lamp.

When the shellac has cooled completely, the slightly raised portion is carefully sanded down level to the surrounding surface. Use #360 wet-or-dry paper on a small narrow block (a cork from a bottle is ideal). Dip the block and paper into rubbing oil and carefully cut down the shellac until it is level with the surface. The dull spot caused by the sanding can be brought back to match the sheen of the surrounding surface using padding lacquer. Blending stains or powders may be intermixed and used in combination with the padding lacquer to blend the color of the patch into the surrounding surface.

"Abrasol", a commercial product used with a felt block may be used to level off a burned in patch with the surrounding area.

When missing pieces of wood, dents or other voids are too large to be filled in with any type of filler, a new piece of wood must be substituted. The damaged area must be cleaned and leveled off, a new piece of wood selected and shaped to fit, then bonded into the prepared area.

PREPARING A SURFACE FOR FINISHING

Any imperfections on a wood surface such as the planer marks on a milled board or any small scratches, even those left by improper sanding, will be magnified and amplified visually by the addition of finishing materials, fillers, stains and finishes. Therefore it is imperative that you do a thorough, careful and complete job of scraping and sanding. Refer to Abrasive chart on page 74 for selection of proper abrasive.

CABINET HAND SCRAPER A simple and most valuable tool for the woodworker. Its sharp, hook shaped edges remove very fine shavings from a wood surface. It is used as the intermediate step between planing and sanding and, when properly used, will eliminate the need for excessive sanding with the coarser grits of sandpaper.

The key to using a scraper is sharpening or shaping the edge of the tool. This will take patience combined with practice in using the scraper, but will prove to be well worth any time you may spend. After these steps are mastered, you will accelerate the time and the quality of your wood surface preparation. With a sharp scraper and good techniques, you can easily eliminate about 75% of any sanding on any wooden surface.

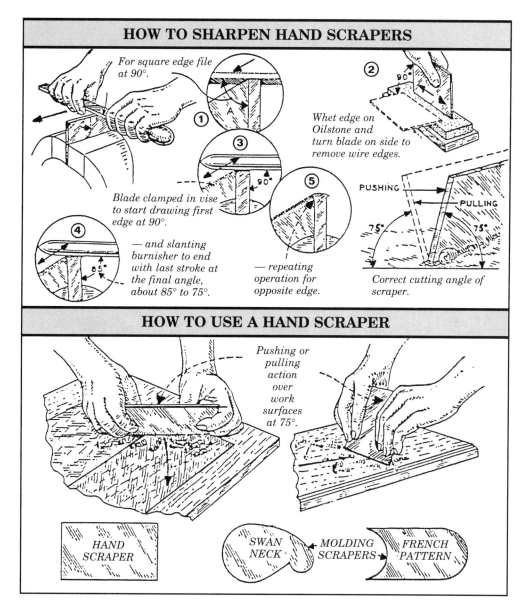

HOW TO SHARPEN HAND SCRAPERS

For square edge file at 90°.

①

Whet edge on Oilstone and turn blade on side to remove wire edges.

③ 90°

Blade clamped in vise to start drawing first edge at 90°.

④ 85°

— and slanting burnisher to end with last stroke at the final angle, about 85° to 75°.

⑤

— repeating operation for opposite edge.

PUSHING PULLING

75° 75°

Correct cutting angle of scraper.

HOW TO USE A HAND SCRAPER

Pushing or pulling action over work surfaces at 75°.

HAND SCRAPER

SWAN NECK ← MOLDING SCRAPERS → FRENCH PATTERN

SANDING After planing and scraping if necessary, a careful, thorough job of sanding is the prerequisite for a beautiful finish. Any unsightly and unwanted marks must now be eliminated with the sandpaper.

If the surface has been properly planed and scraped, you can start sanding with a #180 grit paper, followed by #220 and finally with #240. On flat surfaces, the use of commercial sanding blocks is recommended, or you can wrap your sandpaper around a flat wooden block about 3" x 5". On curved or irregular surfaces, fold small pieces of sandpaper and work the areas carefully by hand.

Often, on moldings, you can wrap your sandpaper around a block of styrofoam and press this into the irregular surface, (the styrofoam will conform to the contour) and complete sanding these irregular surfaces. Abrasive cords and tapes or a Sand-O-Flex should be considered for use on irregular surfaces.

ABRASIVE PAPERS USED IN FINISHING AND REFINISHING

Grit	"0" Series	General Category	Remarks
600	—	Ultra-Fine	Produces High Satin Finish. Used for Wet Sanding.
500	—	Super-Fine	
400	10/0		General range of abrasive papers used for Wet Sanding Lacquer and Varnish Top Coats.
360		Extra-Fine	
320	9/0		
280	8/0	Very-Fine	For Dry Sanding most finishing undercoats. These grades will not show sanding marks.
240	7/0		
220	6/0		
180	5/0	Fine	For Sanding Bare Wood. Next to final grit.
150	4/0		
120	3/0	Medium	For General Wood Sanding. Used prior to selected final grits.
100	2/0		
80	1/0		

Note how the unsanded end grain absorbs too much stain, while the sanded and sealed portion accepts stain evenly.

END GRAIN SANDING Due to the cellular structure of wood, end grains will absorb more finishing materials than the other surfaces, resulting in unsightly, uneven or darkened portions of the wood surface. It is imperative that these end grains be sanded smooth, almost glass like, using the finest grades of the abrasive papers, #240.

Finishing Sander.   *Speed-Bloc Sander.*

POWER SANDERS There are many types of powered sanding equipment available to the woodworker. Some when not properly used may cause severe damage to the surface. Two power sanders that will produce satisfactory results are the "Finishing Sander" or the "Speed-Bloc."

TIPS & TECHNIQUES

1 Prior to sanding with the finest grit sandpaper, dampen a clean cloth with clear water, wring out and rub the wood surface. This will raise those very fine wood "whiskers" which, when the surface is dry, can be removed with the #240 sandpaper.

2 Always sand in the direction of prominent grain.

3 Clean the surface of the sandpaper often. Use either a brush or an abrasive cleaner stick.

4 Prior to the application of any finishing material, wipe entire surface with a tack cloth to remove all sanding dust.

5 Certain types of wood (Teak, Rosewood & Cedar are the most common) have natural oils and waxes on their surface which could repel any finishing material you contemplate applying. To help eliminate this potential problem, use a clean rag dampened with either lacquer thinner or denatured alcohol and wipe the entire surface just prior to the application of your finishing material. It will prove to be good practice to do this on all bare wood surfaces.

SECTION 3

The Finishing Processes

After you have joined, repaired and bleached (if necessary) your wooden surfaces, and have, of course properly sanded them, you are ready to apply the stains, paste wood fillers and finishing coats that will (combined with the beauty of the wood itself) magnify and enhance your conscientious efforts.

The choice of correct finishing materials for your woodworking project is perhaps the most important step in the final finishing or refinishing of a wooden surface. During the past decade, modern chemical research has offered the craftsman new, synthetic chemical compounds that are equal to and even surpass the finishing materials that were manufactured and used in the past. Presently, research continues at a rapid rate and the potential for further advancement in the field of finishing materials appears unlimited.

THE FINISHING PROCESSES

The general recommended procedure for finishing a prepared wood surface is:

1-Stain
2-Fill
3-Seal
4-Finish
5-Hand-rub and polish

This is not a hard-and-fast rule, but rather a general recommendation. Some of these steps can be eliminated or accomplished in a different order. The final selected procedure is up to the finisher.

When one finishing material is placed over another, it **must bond to material below** it either chemically, (reacting with the material; "melting" into it) or mechanically, by grasping to a rough surface (hence the need for sanding between coats for certain materials). Unless you are absolutely sure of the final results in the application of your selected finishing materials, **test the entire schedule out first** on a scrap of the same type of wood the finishing materials will be applied to, or in some inconspicuous area if you are refinishing a fine piece of furniture.

Be aware that some finishing materials are **NOT** compatible with other finishing materials, especially if you alternate between different manufacturers. Too often I have heard of using a stain by one manufacturer and a finish by another, etc. and these surfaces never dry. Only by testing, experience and following printed instructions, can you possibly avoid this very common potential hazard.

SCHEDULES

A finishing schedule is a step-by-step procedural guide. Most home owners have developed their own finishing schedules and most often little more is done than slapping on another coat of gummy furniture polish on top of a dirt embedded filmy surface. This practice is usually the result of not knowing any better revitalizing techniques. To assist you in the rapid evaluation of the application of a finish, schedules are indicated and dispersed throughout the text where applicable.

STAINING:
GIVING COLOR TO WOOD

Stains are applied to a wood surface to emphasize the natural beauty of the grain and color of the wood. They may also be used to blend or harmonize the color of a surface fabricated of several types or multiple pieces of wood. They can be used to camouflage unattractive grains and they can also be used to make an inexpensive piece of wood appear more traditional. They are invaluable in patch and repair work.

Stains are colored materials either natural or synthetic, suspended in a liquid or vehicle. The liquid or vehicle, eventually oxidizes or evaporates out of the wood leaving only the colored material behind. The coloring material is provided either through soluble dyes which penetrate into the wood fibers and alter the natural wood color, or insoluble pigments which remain on the wood surface and partially cover the wood grain and do not penetrate too deep into the wood surface.

Your primary consideration for Stain selection should first be color. Next should be the type of stain you want to use, Water, NGR or Oil (see Section I for a discussion of the various types of stains). A pigmented type or dye type of stain. A penetrating type or a surface type. The method and equipment available for application and possibly the cost and availability of the materials. All these factors should be carefully considered prior to your final selection of a stain.

NOTE: Every stain manufacturer have their own particular nomenclature, i.e.: what one manufacturer calls "Walnut" will NOT necessarily match another manufacturer's "Walnut," etc.

STAINING SOFT WOODS and END GRAINS

Sealed   Unsealed

Due to the natural cellular structure of wood, certain types of wood absorb stain unevenly, leaving a "blotchy" appearance. This condition usually prevails on softwoods such as Pine and Fir, however, it can happen on other woods, hence the precaution of making a test first. The application of a Wash Coat or a Sealer on the wood surface prior to staining will prevent or minimize the unsightly appearance of uneven staining.

End grain, unless properly prepared, will stain much darker than the surface or top grain of a piece of wood. To achieve a uniform stain on the surface and end grains of a piece of wood, a super job of sanding the end grain is a must, followed by the application of one or two coats of a sealer prior to the application of the stain.

An alternate to sealing end grain as mentioned above, is to prepare a paint mixture, using Japan Colors thinned with lacquer thinner. Mixing White, Raw Sienna and Burnt Umber Japan colors, combined with lacquer thinner, will produce a reasonable wood color. Trial and error will dictate the proportions to use to match the color of the wood you are working on. After this "paint" has been applied and dry, you can proceed to stain over this area along with the remainder of the piece of wood and expect a uniform distribution of the color on all surfaces.

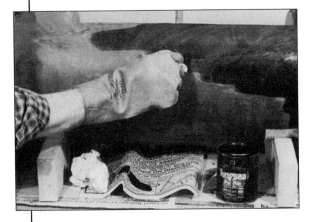

APPLICATION OF STAIN

Stain, the material that, when applied to the wood surface, will enhance the beauty of the wood, will also amplify any imperfections in the surface such as scratches or "swirl" marks from improper sanding. Therefore, the application of the stain must proceed with caution on a properly prepared surface.

The instructions on the container will usually indicate the manufacturer's recommended method of application, by brush, pad, foam applicator, or spraying. Testing and experimenting will soon dictate the method most suited for you.

One sure method of controlling the staining process is as follows:

> Pour a little of your selected stain into a ceramic or glass dish. Dip a pad of cheesecloth into the stain, then "blot" the pad on a clean sheet of paper (avoid paper that has print on it) so that all excess stain has been removed from the pad. Now wipe the wood surface with this stained pad. Repeat as necessary until the entire surface is stained.

Having very little stain on the cheesecloth gives you complete control over the application. You can always make the color darker by repeated applications of the same stain or by selecting a darker stain, but if the applied color is too dark, the only way to lighten it is by bleaching out this unwanted stain, neutralizing, preparing the area once again and restaining.

The following chart may be helpful in trying to match colors.

TO INCREASE:	ADD
Brown Tones	Dark Walnut
Red Tones	Cherry or Red Mahogany
Yellow Tones	Golden Oak
Orange Tones	Colonial Maple

To lighten or reduce the intensity of a color, just add the proper thinner, prior to application of the stain.

TIPS & TECHNIQUES

1 Apply stain sparingly and gradually. Additional coats of stain will usually darken the color or you can use a darker colored stain. When staining, remember, you can always make a color darker but to lighten a color already applied to a surface will require bleaching the surface and restaining to the proper color.

2 Stains can be intermixed, providing they are all in the same family, i.e.; any NGR stains can be mixed together, any Water stains can be intermixed, etc. Do NOT mix water stains with NGR or Oil stains, etc. Stay within the same solvent or vehicle family.

3 When mixing stains—keep a formula for possible future reference.

4 Use only Glass or Ceramic containers to mix stains. When mixing water stains, it is recommended to use distilled or bottled water.

5 With all finishing procedures, if you are not sure of the result of using the various finishing procedures and materials, it will be advantageous to check out the entire finishing schedule on a scrap piece of the same type of wood you are going to work on. When you examine the final result, you can then make any necessary alterations or adjustments to your finishing materials and procedures, prior to working on your final piece.

6 Do NOT attempt any finishing process if the Humidity is above 50%. The ideal temperature range is from 60° to 75°.

7 When all else fails, read instructions.

FILLING

PASTE WOOD FILLER

Filler will fill up the pores or grain in the wood surface and provide you with a glass-smooth surface upon which you can build up the subsequent coats of finishing material. If a stain was used, allow it to dry completely prior to the application of a paste wood filler.

Fillers may be used on open grained woods such as Ash, Mahogany, Chestnut, Walnut, and Oak. Generally, close grained woods such as Cherry, Maple, Birch, and Beech do not require a filler. Colored paste wood fillers may be used to help accent the grain in open grain woods and also stain the surface at the same time.

Select a paste wood filler with the color of your choice or use the natural filler and Japan colors to tint the filler to the color of your liking. Use a clean, empty can and mix the filler with turpentine or quality mineral spirits, to the consistency of a heavy cream. Be sure the mixture is completely mixed, no lumps. Apply the filler to the wood surface using a fairly stiff brush, using a circular motion to work the filler into the pores and grain of the wood. Always keep the prepared, mixed filler in the container well mixed as it has a tendency to settle. After about 20 minutes, the filler will start to loose its "wet" appearance and start to dry to a dull finish.

Apply filler with a brush. Brush first with the grain, then across grain over the first filler coat.

After about 20 minutes the filler will start to dry to a dull finish.

Wiping paste wood filler from surface with burlap.

When the surface has lost its "wet" appearance and is entirely dull, use a piece of burlap or coarse rag such as old toweling and wipe the surface of the wood **across** the grain. Continue to wipe across the grain until all traces of the filler are removed from the surface, except in the pores and grain where it is wanted. Finish wiping with clean rags, **stroking lightly with the grain** until a sheen appears. Inspect the surface to be sure that all pores and grain are filled. If voids are still visible, apply a second coat of filler and repeat the above procedure. A good fill job is 90% of a good finish.

When filling a large surface, keep an eye on the surface that you have already filled. If the filler is allowed to dry too long, it becomes most difficult to remove.

Once the surface is clean and smooth and no residue of the filler is left, allow the piece to dry for at least 24 hours prior to the application of any other finishing materials. It is important that the filler be bone dry before proceeding. The surface should now be **lightly** sanded with a #220 or #240 grit garnet paper. Use a Tack rag and remove all dust and dirt from the surface.

For a first class finish it is suggested that you consider applying a coat of sanding sealer over the surface. This will lessen the chance of the stains and filler "bleeding" into your finishing coats. When the sealer is dry, sand the surface lightly with #240 garnet paper, remove all dust using a tack rag and then proceed to apply your final finish coats of lacquer, varnish or polyurethane.

By using stains and fillers of different colors, such as a dark stain and a light colored filler or visa versa, beautiful two tone work can be achieved. It is difficult, but also possible to match the color of the wood by adjusting the color of these products.

It is not necessary to fill all woods. Pine, Poplar, Basswood and Cedar are exempt from filling. Woods such as Beech, Birch, Cherry, Maple and Redwood, may or may not be filled, most often NOT. Walnut, Mahogany, Ash, Rosewood, Butternut, Chestnut and Oak are examples of woods that should be filled. The final judgement to use a filler or not, is up to the finisher.

TIPS & TECHNIQUES

1 Do NOT allow the filler to remain on the surface much longer than an hour before wiping the excess from the surface. Paste wood filler dries extremely hard and it will be very difficult to remove the excess from the surface if allowed to dry too long.

2 Woods that are usually filled; Ash, Butternut, Chestnut, Elm, Hickory, Locust, Mahogany, Oak, Rosewood, Walnut.

3 Woods that are NOT usually filled; Basswood, Beech, Birch, Cherry, Cypress, Gum, Ebony, Maple, Poplar, Sycamore, Willow, and most any softwood.

4 Paste wood fillers are opaque in color and will obscure any grain or wood color under them.

5 TEST if you are not sure of the final outcome of using a paste wood filler.

6 If a penetrating type finish is to be used as the final finish, it may be mixed with the filler if filling is necessary.

CAUTION!
Paste Wood Fillers are COMBUSTIBLE.

Keep away from heat and open flame. Avoid prolonged contact with skin and breathing of vapors. Use only with adequate ventilation. Tightly close container after each use and dispose of all used rags promptly and properly.

87

SEALING

Sealers are often used, and are applied to a wood surface for several reasons. They act as a catalyst to provide a better bonding surface between the wood and the finishing material that will ultimately be applied upon it. They can also be used to prevent uneven absorption of stains or a finish on certain species of wood, especially on softwoods. Sealers, when applied over a stain, help to eliminate the possibility of the stain "bleeding" through, into the subsequent coats of finish. When used on porous end grains, a sealer will help to prevent the excessive, uneven absorption of a stain. They reduce the tendency of the grain on such woods as Fir and Pine to show through paints and enamels.

Soft woods, from needle-bearing trees such as pine, fir and spruce, may take stain unevenly in addition to absorbing lots of color fast. To avoid a blotchy look, it is advisable to apply Wood Conditioner (sealer) before staining. Maple, although a hard wood, has a tendency to absorb dark finishes unevenly. Pre-treating with Wood Conditioner to avoid a spotted effect is recommended.

There are many prepared sealers available. They are often referred to as "Sanding Sealers, Wood Conditioner," etc. They are basically a special formulation of lacquer or combination of oils and finishing materials. They may be wiped on the surface with a rag, applied with a brush or sprayed on the surface. Sealers dry fast, usually within 2 hours, after which the surface can be lightly sanded with a #240 sandpaper. Do **NOT** flood the sanding sealer on the surface. Two medium coats are far superior to one heavy coat. Allow each coat to dry thoroughly, sand properly, then remove the sanding dust before proceeding to the next step.

Other materials may be used as a sealer. Often it will be a reduced or thinned out formulation of your final finishing material. Follow the manufacturer's instructions for thinning and application.

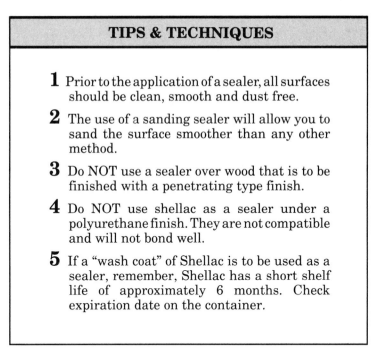

TIPS & TECHNIQUES

1 Prior to the application of a sealer, all surfaces should be clean, smooth and dust free.

2 The use of a sanding sealer will allow you to sand the surface smoother than any other method.

3 Do NOT use a sealer over wood that is to be finished with a penetrating type finish.

4 Do NOT use shellac as a sealer under a polyurethane finish. They are not compatible and will not bond well.

5 If a "wash coat" of Shellac is to be used as a sealer, remember, Shellac has a short shelf life of approximately 6 months. Check expiration date on the container.

FISH EYES

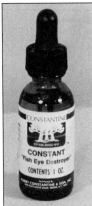

The following sections are prefaced by these few words about a very common ailment, "Fish Eyes," which, if encountered, can cause you much grief and distress. This potential aggravation can easily be eliminated by considering the following simple suggestions.

Fish Eyes may appear on the surface of a finish as small round craters and are the direct result of the presence of silicones on the surface to which you are applying another material or finish.

Silicones are present in most modern strippers and furniture polishes and are the reason for their ease of use. Their presence during a finishing or refinishing operation will result in a disastrous appearance of the finish. If you suspect the surface you are working on is contaminated with silicones, wash the surface with lacquer thinner or a special silicone remover (not Fish Eye Destroyer) sold for this purpose.

Fish Eyes are very hard to remove once the finishing material is dry. If you notice the appearance of these Fish Eyes before the surface is completely dry, remove the "wet" finish with a rag soaked in the appropriate thinner or solvent and start over. Small Fish Eyes can be sanded out and refinished over, but large ones will leave a noticeable depression.

Fish Eye Destroyer is available to help prevent this problem. One dropper full of this material added into one pint of your finishing material will practically eliminate the problems with Fish Eyes. Many professional finishers routinely add the "Fish Eye Destroyer" to their finishing materials to remove one more potential aggravating problem.

HOW TO REMOVE WAX OR SILICONES

1 Remove emulsion or water based waxes or silicones by scrubbing with a solution of 1 cup household ammonia, 1/2 cup heavy duty detergent and three cups of warm water. Use household cleanser over stubborn areas, exercising care not to cut through the finish. Rinse with clean water and dry thoroughly.

2 Remove solvent based waxes or paste wax with turpentine or mineral spirits. Use clean rags and change them frequently; otherwise the wax will just be spread around and not removed.

3 If type of wax is unknown, use procedures 1 and 2 above.

4 Commercially, "De-Waxers" are available for those who wish to purchase and use them.

APPLICATION OF FINAL
TOP PROTECTIVE COATS

No matter whether you apply a single coat or multiple coats, the success or failure of this material will depend upon the type and quality of material selected. Regardless of how much labor and materials were expended on the undercoats, if the top coat fails, your time, labor and materials may be wasted.

The Topcoat has to perform a number of functions:

1 Upon it depends whether the finish has a glossy, flat, or intermediate appearance. This can be accomplished by rubbing and polishing or by selecting a finish of the required luster.

2 It must lend to the appearance of fullness to the finish.

3 It must cover completely, so as to remedy any skips in the undercoats.

4 It must provide both abrasion and moisture resistance and in some cases must be resistant to alcohol, cosmetics and detergents.

5 It must adhere firmly to whatever it is applied to.

Your requirements should dictate what finish you should use. Shellac, Lacquers and Varnishes have been around for years. In addition to these, modern research has now offered us a wide spectrum of finishing materials. Many of these new finishes have a high resistance to the various conditions and materials to which they will be exposed. The methods of application has been simplified, along with the chore of cleaning up. The durability and longevity of these new finishing materials has been greatly increased. Use what will give you the best results over the longest period of time.

There are two basic types of finishing materials:

1 **Penetrating Finish** One that penetrates into the wood surface, polymerizes, (hardens) and dries by oxidation from the inside of the wood up to the surface. A penetrating finish is NOT suggested for use on a surface that has been sealed or filled.

2 **Surface Finish** Finishes that are applied to a surface and dry either by oxidation or evaporation of the solvent.

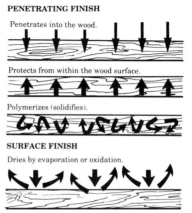

PENETRATING FINISH

Penetrates into the wood.

Protects from within the wood surface.

Polymerizes (solidifies).

SURFACE FINISH

Dries by evaporation or oxidation.

If the materials you use are not compatible, either chemically or mechanically, the materials will separate (peel) or not dry properly. Be sure to read the instructions on the container or TEST.

Only time and experience will allow you to select and use a variety of materials with satisfactory results. Again, if you are not completely sure of the final results, it would be wise to TEST out the entire finishing schedule on a scrap piece of wood, the same as you are going to ultimately finish, or on some inconspicuous spot on the piece of furniture.

When the word "**Varnish**" appears throughout this text, it refers to "**Urethanes or Polyurethanes**," the modern type Varnish, unless otherwise stated.

LACQUER

Lacquer is the finishing material most commonly used today in the furniture manufacturing and custom finishing and refinishing industries. This is due to the rapid drying time of lacquer. Under ideal conditions and with the use of modern applicators, multiple coats can be applied in a very short period of time, and the fast drying characteristic of lacquer inhibits dust from adhering to a wet surface.

Lacquers are available in a wide variety of colors and lusters, from matte to high gloss. They are also available with specific resistance characteristics. Of all the finishing materials, in the clear formulation, lacquer presents the most natural representation of the surface which it covers, be it a stained or a natural wood surface.

The biggest drawback to lacquer is the method of application. Due to its fast drying characteristics, lacquer should be sprayed on the surface. Aerosol containers of lacquer are available and so is spray equipment, but both these are relatively expensive for the home handyman.

LACQUER APPLICATION

Most lacquers are formulated to be applied with quality spray equipment. If the spraying lacquer appears to be slightly thick or heavy, it may be thinned or reduced with lacquer thinner. If multiple coats are to be applied, allow at least fours hours between coats. If sanding is necessary between coats, use a dry, not wet, paper about #400 grit and a Tack cloth to remove dust. If necessary, the final coat, when properly dry, can be hand rubbed or polished to a Satin Finish or a Mirror Finish.

The application of lacquer with a brush may rapidly cause a person to lose their self confidence. Lacquers, by nature, dry very fast, and are very difficult to apply with a brush. Normally for brushing application, you obtain "regular" lacquer which has been manufactured for spray application. Add lacquer "retarder" which slows down the drying time and also add lacquer thinner until you have the consistency that you can successfully brush on a surface without fear of brush marks. For specific amounts to be added, refer to the instructions on the container and make tests. Prepared brushing lacquers are available for purchase.

Do **NOT** attempt to apply lacquer if the air humidity exceeds 50%; "blushing," a white milky appearance, may be observed in your finish.

Do **NOT** apply Lacquer over a Varnish, Wax, or Polyurethane finish. Lacquer may be applied over a sanding sealer.

Do **NOT** try to apply one heavy coat of Lacquer, build up the finish with many thin coats.

POLYURETHANE

Polyurethane is one of the most widely used synthetic type resin "varnishes" used today. It produces a very durable finish, resistant to abrasion, wear, moisture, alcohol. Its luster has great longevity. It may be brushed or sprayed on a surface, and is also available in Aerosol cans. It is supplied in many clear lusters or colored as a stain or paint. Polyurethanes may be applied to new wood surfaces or old surfaces that have been properly prepared for refinishing.

A Satin finish polyurethane gives the wood surface a clear softly polished durable finish which is tough and easily maintained.

A Gloss finish polyurethane gives the wood surface a highly polished glass-like finish. Its deep luster allows the textured beauty of the wood to shine thru. It provides a tough and easily maintained surface.

Applying a polyurethane finish.

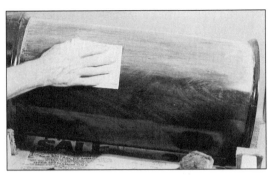

Sand lightly between coats.

POLYURETHANE APPLICATION

Polyurethanes do **not** bond chemically to other finishing materials. The bonding or adhesion to other materials is based solely upon a mechanical bond; therefore, it is mandatory that you do a careful and thorough sanding job on the surface to which it will be applied.

Polyurethanes may be applied with a good bristle brush, foam applicator, or sprayed. They may be thinned with mineral spirits or turpentine. At least two coats are recommended. The more coats the more durable the surface becomes. On surfaces such as desk or table tops that will be subjected to active use, four to six coats is recommended, each coat being allowed to dry properly then lightly sanded and wiped with a tack cloth prior to the application of the subsequent coat.

The final coat should be allowed to dry about a week, to allow it to become hard, if the surface is to be hand rubbed for the ultimate luster.

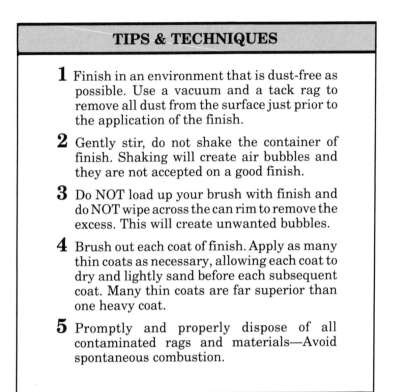

Drying Time

Manufacturers usually specify the suggested drying time for their material. A good general rule of the thumb, if you are not sure of the outcome when applying subsequent coats of finishing materials, is to wait overnight or 24 hours between each step in your finishing schedule. Allow the materials to react and dry properly.

PENETRATING FINISHES

Penetrating finishes, marketed under such names as WATCO, Danish Oil, Tung Oil, Penetrating Oil, etc. are perhaps the easiest finish to apply and have come to be accepted by both the professional and the novice. The material penetrates into the wood and hardens or polymerizes from within the wood up to the top surface of the wood. The finished surface is highly resistant to moisture, stains, heat, minor burns and abrasions.

They are often mixed with dyes or pigments to create a material that colors and finishes in one application. The degree of luster depends on the number of coats applied.

Some manufactures offer a compatible wax that can be applied over the finish when it is dry. They may be purchased in Aerosol cans or regular containers.

Since these finishes dry by oxidation from within the wood to the top, this top surface will be the visible, usable surface when the material is completely dry. Hence the need for a super sanding job which ultimately will offer you a beautiful, smooth protected surface.

PENETRATING FINISH APPLICATION

These finishing materials are easily and usually applied with a rag. They may also be applied with a brush, applicator or sprayed on. A generous amount is applied to the wood surface and allowed to be soaked up from ten to twenty minutes after which time the excess, which will appear "wet," is wiped from the surface. You can repeat the above operation as many times you want to achieve the luster that satisfies you. Variations of this technique might be considered such as sanding with fine sandpaper to form a fine filler just before wiping off the excess material.

Additional coats can be applied at anytime; a day, week, month, or year later, to bring back the luster. Damaged areas are easily repaired by lightly sanding the damage and then simply applying more finish.

Exercise care in applying a penetrating type finish to a veneered surface—too much finishing material could penetrate through the veneers and soften the adhesive causing the veneer to become loose.

SUMMARY

Be sure to read the instructions on the container.

There are literally thousands of manufacturers of finishing materials each having their own specific formulation and instructions. On occasion they do alter their formulas and instructions.

Be aware that research still continues and new innovations continue to be offered such as "**Ultra Violet Screen,**" a chemical added to the formulation which helps to prevent "yellowing" of the finish. New formulations offer polyurethanes and lacquers that are thinned or reduced and readily cleaned up after application, using water. Recently, stains and finishes have been combined with a Gel type vehicle which is easily applied with a cloth.

Research continues to produce finishing material with varying degrees of resistance to moisture, chemicals, liquids, heat and abrasion.

MOST FINISHING MATERIALS CONTAIN PETROLEUM DISTILLATES. KEEP OUT OF REACH OF CHILDREN! Keep away from heat, sparks, and open flames. Use only with adequate ventilation. Avoid prolonged contact with skin and breathing of vapors or spray mist. Use of a proper type filter mask or respirator is recommended. Close container after each use. If swallowed, do NOT induce vomiting:

CALL PHYSICIAN IMMEDIATELY !

TYPES of APPLICATORS

BRUSHES

Sometime during the finishing or refinishing process, you will encounter the need for using a brush. The selection of the proper brush for the specific task is prerequisite for quality results.

The primary consideration should be what kind of "bristle" you will need for a specific application. Soft brass wire bristles are ideal for removing old finishes during the stripping operation.

Foam Applicators   Bristle Brushes

A cheap fibre bristle brush is the best to apply the "stripper" with.

A fibre bristle brush is also good for applying paste wood filler to a surface.

Use an Ox Hair or a good soft hair brush for the application of Brushing Lacquer, Polyurethanes and other similar type finishes.

Nylon type bristles may be used to apply water based materials. Water will tend to make natural bristle too soft.

Good brushes are expensive, but given tender loving care, they will serve you well for many years. Keep them clean and stored properly.

DISPOSABLE FOAM TYPE APPLICATORS

These wedge shaped foam "brushes" can be used to apply stains, penetrating type finishes, varnish, and polyurethanes. They are ideal for touch up work. They leave no ugly brush marks or loose bristles. They are available in a variety of sizes or you can easily make your own by cutting a piece of foam rubber to shape and adding a pinch type clothespin for the handle. Do **NOT** use for shellac or lacquers. Dispose of properly when finished.

AEROSOL SPRAY CONTAINERS

Most every color and type of stain and finish is now available in the Aerosol Spray container. They are ideal for small projects or touch up work; however they are expensive and cannot be intermixed if necessary. Empty containers must be disposed of properly.

The following helpful hints on the use of aerosol products are presented for your consideration.

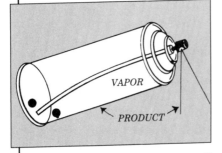

Figure A

Note: Some aerosols require shaking before use, others do not. Some are held upright, others not. This depends upon the product. It will be explained on the label, so read it before you press the button.

Shaking of the can prior to use is very important, particularly with a pigmented material as it is necessary to free the steel ball agitators and mix the contents thoroughly for best results.

Most aerosol products are manufactured with a mark on the top of the valve cup (fig.A) near the spray head indicating the direction in which the head should be pointing to completely empty the can of its contents.

This mark is also located on the same side of the can as the bottom of the dip tube. If this mark should accidentally be removed, depress the spray head while the can is almost horizontal; when the material flow is steady without hesitation, that will indicate the location of the dip tube. (fig. B)

While spraying an object it is best of course, to have it laying flat. This spraying should be done in a well ventilated area with surrounding objects covered.

When spraying, the best method is to start from the front (side nearest to you) and spray back and forth towards the rear, overlapping each stroke. (fig. C) This method basically does two things: number one, eliminates over spraying on the piece and number two, avoids a build up or puddling of material on the ends. **Best results are obtained with multiple light coats rather than one heavy coating.**

As important as cleaning a paint brush after use is the cleaning of an aerosol paint product.

Turning the can upside down and depressing the spray head for a few seconds or until only vapor comes

Figure B

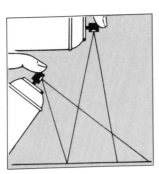

Figure C

out (fig. D) clears out not only the small spray head itself, but also clears the dip tube of paint which could possibly seep and clog the valve on spray head.

If a spray head does become clogged, it can be cleaned by either soaking it in solvent for a period of time or by using a thin knife or similar object to clean the small slot in the bottom side of the head. A pin may be used to clean the small orifice in front of the head. (fig. E)

Aerosol products are often filled with a specific valve and sprayhead which in some cases are not interchangeable from one to another because of the size of the valve opening. Each product is filled with one of several hundred variation of size of slots and orifices for best spray patterns and finish results.

No matter what type of propellant is used...a fluorocarbon, hydrocarbon or other...the propellant is safe when the product is used according to directions. The type of propellant in ratio to the amount of product and in relationship to the design of the valve determines the size of the particles sprayed, which may be from a fine mist to a steady stream.

Read the Label

When you purchase an aerosol product, get acquainted with it. Look it over, read the label. The manufacturer not only provides information for proper use and safety, but very often gives you tips and suggestions by which you will obtain the best and most economical results from the product.

If you treat an acquaintance well, he may turn out to be a friend. If you abuse him, he may become an enemy. This is true also of the things you use every day... a safety pin, an electric light, an automobile, or an aerosol.

All aerosol product instructions caution you to keep the product out of the reach of children. Youngsters often do not understand the consequences of an act. They frequently misuse or abuse things, including their toys. That is why both government and industry establish methods, standards, and devices which seek to protect them. But you have a role in this, too. Either teach the child how to use household things correctly and properly so it won't get hurt, or keep them high out of the child's reach.

Courtesy of Star Finishing Products.

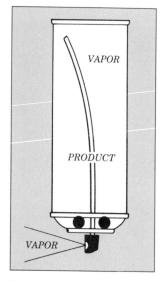

Figure D

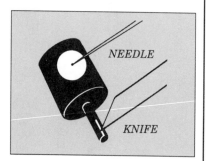

Figure E

99

SPRAY EQUIPMENT

There are three basic types of spray equipment available for applying finishing materials: Conventional, Airless, and Turbo. The use of any type spray equipment requires some experience, testing and experimenting with the material you are going to apply, most importantly with the viscosity.

Conventional The original system, this consists of a compressor which compresses air and delivers it at an adjustable, variable pressure from 20 p.s.i. to 100 p.s.i. (which atomizes the material) at a low volume of approximately 7cfm. Most finishing materials can be sprayed with this equipment providing it is finely adjusted; however, a certain amount of overspray will result. The compressor is usually heavy and bulky. The surrounding air, which can contain dirt and moisture, is inhaled into the compressor through the intake, compressed and then forced out through the gun with the finishing material. To eliminate these contaminates, oil, water and dirt separators or filters should be added to the compressor system and the filters constantly cleaned.

Airless This is essentially a high pressure fluid pump which forces the material through a small orifice at the end of the gun with a pressure of 2000-3000 p.s.i. No air is used. The material can be sprayed thick and without reducing. The material is literally blasted out of the gun, resulting in a large amount of overspray and bounceback. For this reason, this type of equipment is not recommended for fine finishing.

Turbo Spraying (High Volume, Low Pressure) This system is about twenty years old and recently has had a great impact on the application of finishing materials. The power source is a Turbine, not a Compressor. It works on the principle of delivering and using a high volume of air, 40cfm-90+ cfm (depending upon unit selected) at a low pressure of only 3-4 p.s.i. The material is atomized by the large volume airflow and not high pressure, resulting in a negligible amount of overspray which in turn results in a savings on the amount of materials used. 80+% of the material that leaves the nozzle of the gun is transferred to the intended surface. The turbine (think of it as jet aircraft engine), while generating and delivering a large volume of air, is also heating this air sufficiently to dry out moisture that could be absorbed through the intake. Dirt and dust particles in the air are filtered out with a simple, easily cleaned filter.

The Apollo Sprayers International has been a leader and pioneer in the development and promotion of this system which in the face of immediate and forthcoming environmental laws and regulations is the system to be considered. A variety of units are available for the hobbyist up through the most professional requirements and all units are easily portable.

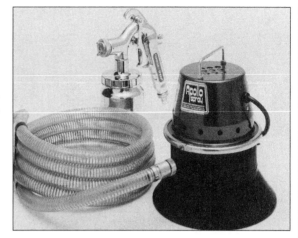

Model 500 Apollo Sprayer

p.s.i. = pounds per square inch;
cfm = cubic feet per minute
HVLP = High Volume, Low Pressure

THE HAND RUBBED FINISH

For the ultimate shine or sheen to your finish, consider using either Pumice Stone or Rottenstone or a combination of both to rub your finish to a shine or sheen that will greatly enhance the appearance of the finish. It is hard to find a finish that will compare to the beauty of a hand-rubbed finish. The traditional method of hand-rubbing with these fine abrasives has produced superior results for centuries. It is a laborious and time consuming technique, and all your efforts will be rewarded with the beautiful end results.

Practically all types of finishes, shellac, lacquers, polyurethanes and varnishes can be hand rubbed. It is essential however that you have a sufficient amount of finishing material on the surface and that it is completely hard and dry. It is recommended that you allow the final finish coat to dry 5-7 days prior to starting the hand rubbed procedure.

Use the following steps to produce your traditional hand rubbed finish.

Rubbing with finishing paper.

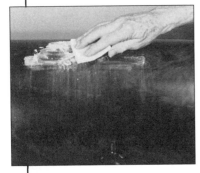

Rubbing with pumice and rubbing oil.

1 After your top coat is completely dry, sand it smooth with wet-or-dry finishing paper to remove any nibs of dust or irregularities in the surface. Start with a 400 grit paper, follow with a 600 grit for the final sanding. You may use either water or Non-Blooming Rubbing Oil to lubricate the paper. To check your progress, occasionally wipe the surface clean, using a clean, soft cloth. After sanding, the surface should have a uniform, dull, matte appearance. Do **NOT** use water as a lubricant if you are rubbing a Shellac Finish. Avoid sanding through the finish.

2 Secure a felt pad, approximately 1/4" x 3" x 5" is the ideal size for most work. Apply a liberal amount of your lubricant, either water or Non-Blooming Rubbing Oil, to the felt pad. Sprinkle some 4F Pumice powder on the finish surface and, using the lubricated pad, rub back and forth over the surface following the grain direction of the wood. The Pumice becomes finer as it wears during the rubbing process. Occasionally wipe away the film of pumice and lubricant to check the progress of your work. When all marks left by your sandpaper or finishing paper are removed or erased, you are done. Wipe the entire surface with a clean, soft cloth and you will see a beautiful satin finish with which you can be satisfied and proud.

3 If you want a higher gloss or sheen, repeat the above step (2), but substitute Rottenstone for the Pumice Stone and use a new, clean felt pad. Periodically wipe away all traces of the Rottenstone and lubricant and see how your shine is progressing. When you have reached the shine or sheen you want, you are finished with the rubbing portion of your job.

4 After you have completed all the rubbing, use a clean, soft cloth and carefully wipe away all traces of Pumice Stone, Rottenstone and lubricant. For the final touch, apply some OZ polish to a clean, soft cloth and wipe over the entire surface. Then with another cloth, wipe and polish the surface for the ultimate shine.

Wiping with OZ polish.

TIPS & TECHNIQUES

1 Whatever lubricant you start with, either water or Non-Blooming Rubbing Oil, stay with it throughout the entire process. Don't change lubricants half way through the process.

2 If you use water as a lubricant, the cutting action of the abrasives will be faster. Using Rubbing Oil as a lubricant may be a little slower, but you will have more control in the progress of the cutting action of the abrasive and not suddenly find that you have cut or penetrated through the finish.

3 Fingermarks, normal dust, dirt and grime can usually and easily be removed by just wiping the surface with OZ polish and lightly polishing. Periodic wiping of a finished surface with OZ polish will enhance the beauty of the piece and will not harm the finished surface or the wood.

4 If you have superficial scratches or white water marks in an existing finish, follow steps 1,2,3 and in most instances they will be removed.

5 If you are rubbing a shellac finish, DO NOT use water as a lubricant.

6 For small areas requiring rubbing, Polishing Compound may be substituted for the pumice stone and rottenstone.

HOW TO FINISH FURNITURE

1) *Proper sanding is important to insure a professional looking finish. To provide a smooth, uniform surface, sand with successively finer grades of sandpaper.*
If the area you are sanding is large, a power sander will speed your work. For smaller jobs, use a sanding block, and be sure to sand in the direction of the grain. Tip: To test for rough spots after sanding, put a sock on your hand and rub the surface of the wood. Sand again where you hit a snag.

2) *Even difficult to reach places should be well sanded for a uniform finish. To sand the cut-out design on the quilt rack, a piece of sandpaper was rolled up to make the job easier. For sanding the round top rail a strip of sandpaper was adhered to tape and pulled back and forth, much like dental floss.*

3) *The traditional method for finishing wood, after sanding, is the application of a stain, to be followed, when dry, by a top coat. Wood Finish by Minwax stains and seals wood surfaces. It penetrates deep into wood fibers so surface scratches will not expose bare wood. A liberal coat of the finish should be applied and allowed 5-15 minutes for penetration before wiping off excess. For a deeper color a second coat may be applied the next day. Wood Finish is available in various wood tone colors ranging from natural to ebony.*

4) The top coat selected for this quilt and blanket rack was a semi-gloss polyurethane by Minwax. It offers durable protection and exceptionally long lasting beauty with the degree of gloss desired. It is also offered in a clear gloss and satin finish.

5) This attractive quilt/blanket rack is an example of some of the diverse and decorative types of unfinished furniture available today. As in this case, unfinished furniture is appealing because it is usually solid wood of high quality, and much less expensive than its finished furniture store counterparts. In fact, browsing through unfinished furniture stores will reveal items which are difficult to find elsewhere.

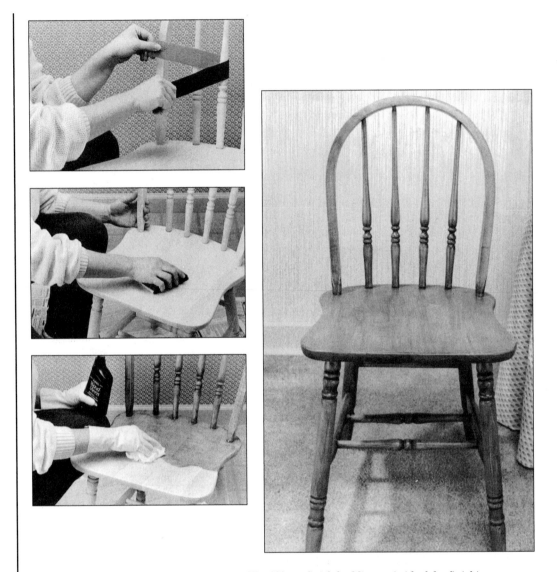

The new one-step WoodSheen finish by Minwax is ideal for finishing a country-style chair. Its hand-rubbed look is appropriate for the traditional style, while its thick consistency permits application with a lint-free cloth, making even spindles easy to finish. Before applying WoodSheen penetrating stain and protective finish, make sure the wood is well sanded. Sand in the direction of the grain with successively finer grades of sandpaper, using a sanding block for flat surfaces. For hard-to-reach places, such as spindles, adhere sandpaper to strips of duct tape and work back and forth much like dental floss. The innovative squeeze bottle makes WoodSheen easy to apply with a cloth without dripping or spilling. Simply wipe it on and in two hours it's dry. Additional coats deepen the color and sheen.

QUICK REFERENCE GUIDE

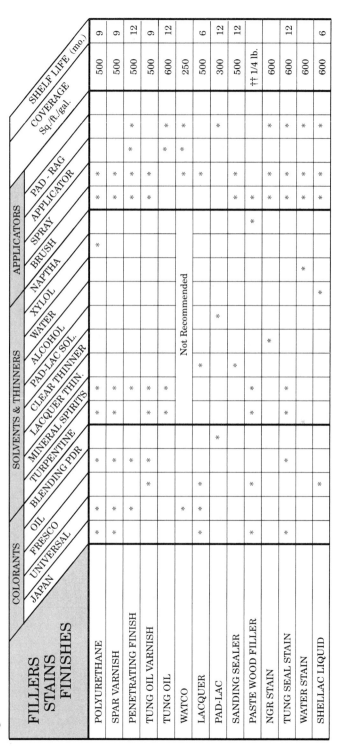

FILLERS / STAINS / FINISHES	COLORANTS				SOLVENTS & THINNERS										APPLICATORS				COVERAGE Sq./ft./gal.	SHELF LIFE (mo.)
	JAPAN	UNIVERSAL	FRESCO	OIL	BLENDING PDR	TURPENTINE	MINERAL SPIRITS	LACQUER THIN.	CLEAR THINNER	PAD-LAC SOL.	ALCOHOL	WATER	XYLOL	NAPTHA	BRUSH	SPRAY	APPLICATOR	PAD - RAG		
POLYURETHANE	*	*	*	*		*	*		*						*		*		500	9
SPAR VARNISH	*	*	*	*		*	*		*						*		*		500	9
PENETRATING FINISH		*	*			*	*		*								*	*	500	12
TUNG OIL VARNISH				*		*	*		*								*	*	500	9
TUNG OIL		*				*	*		*								*	*	600	12
WATCO		*			Not Recommended											*	*	*	250	
LACQUER	*	*	*	*				*		*							*	*	500	6
PAD-LAC					*								*						300	12
SANDING SEALER								*			*						*	*	500	12
PASTE WOOD FILLER	*			*		*	*		*							*	*		†† 1/4 lb.	
NGR STAIN											*	*					*	*	600	
TUNG SEAL STAIN	*			*		*	*		*								*	*	600	12
WATER STAIN														*			*	*	600	
SHELLAC LIQUID			*								*						*	*	600	6

†† 1/4 lb. per sq. ft.

FINISHING SCHEDULES

Schedule #1   **FINISHING OPEN GRAIN WOODS**

1) Apply wash coat of lacquer or sanding sealer. Allow to dry.
2) Sand with #240 Garnet paper. Remove dust with a tack rag.
3) Apply paste wood filler of desired color. Wipe off excess. Allow to dry thoroughly (24 hours).
4) Apply sanding sealer. Allow to dry.
5) Sand with #240 Garnet paper. Remove dust with a tack rag.
6) Apply several coats of lacquer. Allow to dry between coats. **Or** apply coats of Varnish or Polyurethane. Allow to dry.
7) Rub with polishing compound or use pumice and rottenstone. See page 101.

Schedule #2   **FINISHING CLOSE GRAIN WOODS**

1) Apply NGR stain. Allow to dry.
2) Apply sanding sealer. Allow to dry.
3) Sand with #240 Garnet paper. Remove all dust with a tack cloth.
4) Apply several coats of clear lacquer. Sand between coats with #600 wet or dry silicon paper.
5) Allow last coat to dry overnight.
6) Rub with polishing compound or pumice and rottenstone. See page 101.

Schedule #3   **SINGLE STAINING FOR OPEN GRAIN WOODS**

1) Apply NGR Stain. Dry 1 hr.
2) Apply Wash coat of lacquer or white shellac. Dry 20 min. Sand with 7/0 garnet paper. Remove dust thoroughly.
3) Fill grain with paste wood filler in color of your choice. Dry 24 hrs.
4) Apply Sanding Sealer. Dry 2 hours. Sand with 7/0 Garnet paper. Dust thoroughly.
5) Apply second coat of Sanding Sealer. When dry, sand and dust.

6) Apply two coats of Lacquer or one coat of varnish. Allow lacquer to dry overnight, varnish, dry for 4 days.

7) Rub satin smooth with pumice and rottenstone or use polishing compound.

8) Dust and wipe clean.

Schedule #4 — DOUBLE STAINING for OPEN GRAIN WOODS

1) Apply NGR Stain. Dry 1 hour.
2) Apply selected paste wood filler. Dry 24 hrs.
3) Apply sanding sealer. Dry 2 hour. Sand with 7/0 Garnet paper-Remove dust.
4) Apply second coat of Sanding Sealer. Repeat (3).
5) Apply pigmented wiping stain a shade darker than the NGR stain used. Carefully wipe with a clean cloth after 5-10 minutes to achieve desired color. Allow to dry 24 hours.
6) Apply finishing coats of lacquer or varnish. Allow to dry properly. Rub to desired sheen. Dust & clean.

Schedule #5 — LACQUER FINISH for CLOSE GRAINED WOODS

1) Apply NGR stain. Dry 1 hour.
2) Apply sanding sealer. Dry 1 hour. Sand with 7/0. Dust thoroughly.
3) Apply 2 or 3 coats of clear gloss lacquer. Sand between coats with 7/0 or finer paper. Allow surface to dry at least 24 hours.
4) Rub to desired sheen. Dust and clean.

Schedule #6 — DOUBLE STAINING—CLOSE GRAIN WOODS

1) Apply NGR Stain. Dry 1 hour. Lightly sand with 7/0 Garnet paper.
2) Apply pigmented wiping stain, a shade darker than the NGR stain. Dry 2 hours.
3) Apply Sanding Sealer. Dry 1 hour. Sand with 7/0 Garnet paper. Dust thoroughly.
4) Apply two coats of clear lacquer or one coat of varnish. Allow to dry thoroughly. Rub to desired sheen. Wipe clean.

PICKLED PINE WITH WHITE OVERCAST

1) Apply Gray NGR Stain.
2) Apply Sanding sealer or a 2-lb. cut of white shellac.
3) Apply coat of pigmented white wiping stain. This may be wiped clean, left streaked or wiped across the grain to produce the "smoked" effect. Allow to dry at least 24 hours.
4) Sand lightly with 7/0 Garnet paper. Dust and clean thoroughly.
5) Apply 2-lb. cut of white shellac, lacquer or varnish. Allow to dry thoroughly.
6) Rub finish to desired sheen. Dust and clean.

Schedule #8 **PICKLED OR LIMED FINISH**

This type of finish is reserved for open grain woods only.

1) The grain or pores of the wood are filled using either heavy white paint or white paste wood filler. Apply the mixture and remove as you would any paste wood filler.
2) When filler is dry (at least 24 hours) apply a washcoat of your final finish or a sanding sealer.
3) When washcoat or sealer coats are dry, lightly sand and remove dust.
4) Apply sufficient number of final coats and rub or polish final coat if desired. There are many variations of this style of finish.

CARE and MAINTENANCE

After successfully completing all the steps in your finishing schedule, you are now ready to place the piece of furniture into service and have it admired and become subjected to normal use and possible abuse. It is most likely that your finish will be the part that should also have some additional protection and occasional care.

The proper application of a quality paste wax will provide a longer lasting and more pleasing appeal than many commercial type furniture polishes. The wax polish can easily be revived if necessary by gently rubbing with a soft cloth. Additional coats of wax should only be applied when necessary, which will be indicated by the surface refusing to polish after rubbing with a soft dry cloth. Avoid excessive build up of wax as this will greatly alter the appearance of the surface and may also cause the surface to become sticky.

Many commercial type furniture polishes are available. Follow the instructions on the container for the best results. Many of these commercial polishes contain silicones which make for the ease of its use and the protection it provides. However if a surface that has been treated with such a polish ever has to be refinished, these silicone particles can and will cause "fish eyes" in the new finish. Such a reaction in a new finish can be reduced or eliminated by the addition of a product referred to as "fish eye destroyer" into the new finishing material.

Recently, one of the largest finishing material manufacturers introduced on the market a product called "OZ Cream Polish." It cleans, polishes and protects the surface in one easy application, and it does not contain any silicones.

TIPS & TECHNIQUES

1 Dust furniture regularly with a clean soft cloth. Lift all objects on the surface when dusting and dust in the direction of the grain. Use protective pads under objects that may produce moisture such as drinks, flower pots, etc. Also use protective pads under utensils that can produce heat such as hot bowls, platters, etc.

2 If necessary to wash a finished surface, mix a mild detergent or a soap such as Murphy's Oil soap in some warm water. With a soft cloth, gently wash a small area at a time. Rinse cloth and surface often with clean warm water. When the surfaces have dried thoroughly, they may be waxed or polished.

3 Do NOT apply wax in thick coats. Excessive wax will deter from the appearance and make the surface very slippery. Under damp or humid conditions the wax will become sticky.

4 Apply wax and polishes sparingly and only when needed, which will be indicated by the material presently on the surface refusing to "buff or polish" up.

REPAIR and REFINISHING
of an Existing Finish

All prior information and discussion pertained to finishing an exposed raw wood surface. It may have been purchased as an "unpainted" piece, custom constructed or may be a piece that had been completely stripped of all old finishing materials on its surface.

Recent years have seen the introduction of new finishing materials designed to eliminate the messy job of stripping. They are often referred to as "Refinishers, Rejuvenators or Amalgamators." Each manufacturer produces their own particular material, which must be used according to their specific instructions.

Many of these refinishing materials don't totally remove the finish, they merely soften the existing surface, allowing you to level it out. When dry, the result is a level, finished surface now free from the previous crazing, alligatoring, etc. Others first soften or dissolve the old surface, you wipe it off, apply a stain and or a sealer, then apply a final coat. Generally these procedures are easily accomplished using wiping rags only. In summary, this is the general, very basic technique used and will vary between manufacturers according to their instructions.

It is true, they are a less messy, require less time and effort but they are **NOT** a solution to quality refinishing.

The complete removal of an old finish is not always necessary and in the case of authentic valuable antiques may not be warranted. Very often if the old finish is just dull and dirty, a careful cleaning and polishing as described earlier will bring back the beauty to the surface of the piece.

If the original finish is found to be in good condition, not peeling or chipped, it can be revived or rejuvenated after it has been cleaned free of dirt, grease, polishes and waxes. First the type of material used for the old finish should be determined.

Testing the old finish After you have determined that you are going to refurbish or revive an existing finish, find out what type of finish the old finish was. First put a drop of denatured alcohol on some inconspicuous area, if the alcohol evaporates apply another drop. After a few minutes, if the old finish was shellac, it will soften. If it does not soften, repeat the same operation using lacquer thinner this time; if this softens the old surface, it is a lacquer finish. If neither the denatured alcohol nor the lacquer thinner softens the surface, you can be pretty sure the finish is a varnish or polyurethane.

The effort involved in reviving an old finish, whether in its entirety or only a local spot, is well worth trying first.

The following refinishing schedules may be used to revive various types of old finishes.

MISCELLANEOUS SCHEDULES

Schedule #9   **REVIVE A CLEAR OLD FINISH**

1) Thoroughly clean the surface of all old dirt, waxes, grease and polishes.
2) Lightly sand surface with #320 sandpaper to provide a "tooth" for the new material to adhere to.
3) Apply a single coat of the same material that was used originally.
4) When thoroughly dry, hand rub or polish this newly finished surface to the desired sheen.

Schedule #10   **BRIGHTENING OLD VARNISH OR LACQUER FINISHES**

1) Clean with mild soap such as Murphy's Oil Soap. Remove sludge with clean damp cloth. Dry with clean dry cloth.
2) If a heavily waxed surface remains sticky, apply denatured alcohol with a clean soft cloth. Wipe off immediately with a soft cloth. Alcohol left too long could damage the finish.
3) Apply polishing compound (a slightly abrasive paste) on cheesecloth pad. Rub until a high gloss appears uniformly over the surface.
4) Clean up with OZ cream polish.

Schedule #11	**CLEANING and REJUVENATING A WOOD FINISH**

1) Refer to "Washing & Cleaning" in Section 1. Work quickly and don't get the piece too wet. When water remains on a surface too long, a white haze sometimes referred to as "blushing" may develop. This is especially true of shellac and lacquer finishes. Once the dirt has been removed, you may find that the finish has an accumulation of waxes or polishes on it. These may be removed by dipping a clean cloth in a dish of turpentine or De-Waxer. Wring out the cloth and clean a small area at a time. Follow this by gently rubbing with another clean, dry cloth. Continue this process until the entire piece is free from all old waxes and polishes.

2) Mix three parts of boiled linseed oil and one part of turpentine in a clean glass jar that can be tightly covered. Shake the mixture well (this mixture can be stored for long periods of time). Next, pour some hot water into a glass or ceramic dish, then pour some of your solution onto the surface of the hot water. Do not stir the mixture into the hot water, just float it on the surface.

3) Dip a clean, soft cloth into the floating oil mixture, wring out the cloth and commence to rub a small area at a time. Do not get too much moisture on joints or glued areas as the moisture can cause the glue to soften and release its hold. Use a toothbrush on carved or grooved areas.

4) Dip a clean, soft cloth into some clear, warm water, wring out the cloth and wipe the surface that you applied your solution to. Follow this by wiping with a clean, dry, soft cloth. When the water becomes cool, discard the water and solution floating on the top. As the water cools, the solution will become "gummy" and this can cause you more trouble than it's worth.

5) When the entire piece of furniture has been completely washed, dewaxed and the finish rejuvenated, you may elect to apply a good furniture wax or polish such as OZ polish to restore the sheen or shine.

Schedule #12	**REVITALIZING FURNITURE PERIODICALLY**

1) Clean carefully with denatured alcohol. Remove immediately with soft cloth. Work lightly and quickly.

2) Apply OZ cream polish with clean lintless cloth. Pour polish on cloth. Use sparingly. Rub thoroughly but not too hard.

Schedule #13

HIDING SCRATCHES-SIMPLIFIED

For a quick fix on kid's room furniture constantly being subjected to hard knocks, fill small scratches with matching colored touch-up crayons or putty sticks, sold at most finishing supply counters.

Schedule #14

HIDING SCRATCHES and MINOR BLEMISHES

1) Clean with mild soap such as Murphy's Oil Soap. Work over entire surface where blemishes occur. Treat an entire panel if practicable. Remove soap residue with a clean damp cloth.

2) Clean off wax, grease, furniture polish with denatured alcohol. Use alcohol sparingly and work quickly so as not to soften the existing finish. Wipe dry immediately. If you are working on a shellac surface, use turpentine instead of alcohol.

3) Use a small, pointed touch-up brush to apply blending stain mixed with Pad-Lac to blemished area only. Allow to dry 12 hours. A "Q" tip can be substituted for a small touch up brush.

4) Apply polishing compound on a cheesecloth pad. Rub repaired area to restore gloss.

5) If a satin finish is preferred, go over entire surface with pumice stone and rubbing oil on cheesecloth pad.

Schedule #15

SPOT BLEACHING

1) To reduce the prominence of a stubborn dark spot on a lighter surface, bleach the spot.

2) Clorox, a strong household laundry bleach will lighten many such spots. Use a small touch-up brush to apply the Clorox, or a "Q" tip may be substituted for the brush. Let the solution stand for a minute or two. Sop off with a clean damp towel. Repeat several times if necessary, allowing the bleach to stand somewhat longer each time. Be sure to neutralize the area. Wash away all traces of the bleach. Let spot dry overnight. Touch up as in schedule #16 starting with step "3."

Schedule #16 **HIDING FINE CHECKS IN VARNISH SURFACE**

1) Clean with damp (not wet) application of Murphy's Oil Soap applied on a cheesecloth pad. Use somewhat stronger mix than suggested on container for general purposes.
2) Clean off sludge with damp towels. Wipe dry.
3) With a cheesecloth pad, work over surface quickly and lightly in circular motion with two parts Pad-Lac solvent mixed with one part Pad-Lac. Allow to dry 24 hours before using.

Schedule #17 **REMOVING BURN MARKS**

Use a craft knife to carefully scrape away all charred wood. Then repair with shellac sticks as outlined on page 117.

TIPS & TECHNIQUES

1 As with any finishing or refinishing project, unless you are absolutely sure of what the final result will look like, use a small inconspicuous area and perform the entire procedure on this small area. This way you can make any adjustments to your procedure, if necessary.

2 Do not use any finishing or refinishing materials in an area that could cause a fire. Most of the materials are flammable and/or combustible. Dispose of all such materials properly. Provide adequate ventilation.

3 Protect your eyes and skin from hot and harsh solutions.

4 Be sure to read and adhere to all instructions printed on the containers of purchased materials.

5 When using alcohol for any reason on a shellac finish, be aware that alcohol can soften shellac. Use alcohol sparingly, work quickly and cautiously.

REPAIRING SMALL HOLES
using SHELLAC STICKS

1) Equipment: steel spatula, alcohol lamp or can of sterno and stick shellac. When two colors of shellac sticks are close to the color of the finish, choose the lighter color. The finishing procedure will darken it.

2) Clean entire section with one-to-one mixture of denatured alcohol and water. Wash quickly. Don't let solution or pad stand on surface.

3) Heat spatula over soot-free flame or use an electric heated knife. Touch heated knife or spatula to shellac stick. Drip molten shellac into damaged area.

4) Repeat molten shellac step. Press shellac into blemish with hot spatula. Apply shellac until it is slightly mounded above surface. Let dry hard.

5) Cut a piece of felt block 1/4" x 1"x 1". Fold 400A wet-or-dry silicon carbide paper over the 1/4" edge of felt. Dip block and paper into rubbing oil. Cut down mounded shellac until flush.

6) Restore polish to cleaned section with Pad-Lac applied with cheesecloth pad, adding proper color blending powder if necessary.

7) For ultra-high polish, if desired, go over surface with Pad-Lac thinned liberally with Pad-Lac solvent.

8) For the ultimate polish, go over surface quickly and lightly with pad lightly dampened with Pad-Lac solvent.

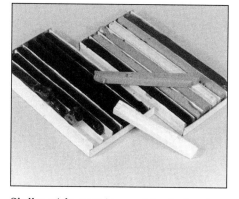

Shellac sticks come in a variety of colors to match any finish.

Restoring polish with Pad-Lac and blending powder.

117

REPAIRING LARGE HOLES

This technique is appropriate to everyday furniture as distinguished from valuable antiques which would be surface-patched with new veneer or solid wood.

1) Clean entire surface with damp application of Murphys Oil Soap. Clean off residue. Wipe dry.

2) Go over entire surface lightly with turpentine to remove wax and grease. Wipe off immediately.

3) Fill crevice with Carpenter's wood filler. Apply as it comes from the container. Pack and mound the area with a steel spatula.

4) When completely dry, sand the area flat.

5) Within one hour after sanding, use small brush to apply matching blending stain dissolved in Pad-Lac. Choose somewhat lighter shade of stain. The finishing procedure will darken it. Dry overnight.

6) If needed, rub treated surface with 400A wet or dry paper held over a small square of felt and dipped in rubbing oil. When thoroughly dry, dust with tack rag.

7) Go over entire surface with Pad-Lac on a pad made only slightly damp by tapping off the solution on a brown paper bag. Slide pad gently on and off surface. Keep in motion, making a figure "8" design.

8) Color-blend the patch by padding, in motion, blending stain picked up on Pad-Lac moistened cheesecloth.

9) Go over the entire section using two parts Pad-Lac thinned with one part Pad-Lac solvent. Allow to dry several days before using.

FRENCH POLISHING

This is a method of applying a resinous finishing material to a surface using a pad. The exact origin of this art is unknown but many old pieces of furniture and art treasures have been preserved by this finishing method.

Old time French polishers used their own formula of shellac flakes dissolved in alcohol and often added small amounts of turpentine and linseed oil, the exact formula being a closely guarded secret of the polisher. This material was applied to the surface with a pad made of a one foot square of linen or lintless cotton cloth, folded twice to form a six inch square, into which a ball of muslin or cotton was inserted and enveloped tightly by twisting the overhanging edges of the line. This was often referred to as the "rubber" or polishing pad.

A typical schedule for French polishing is as follows:

1) All surfaces to be finished must be sanded as smooth as possible and the surface wiped with a rag dampened with water to raise the excess grain. When dry, all surfaces are again resanded using fine sandpaper.

2) Prepare your mixture of shellac or shellac flakes and alcohol plus other materials you may elect to add.

3) Apply your mixture to the pad and disperse this mixture throughout the pad by tapping the moistened pad into the open palm of your hand. This will avoid any wet spots on the pad and help produce an even flow during application.

4) At all times when the pad is in contact with the surface it must be in motion. Use either a pendulum motion, or circular motion. Stopping the pad on the surface will result in cloth marks from the pad being left on the surface. Use your "charged" pad and apply your first application evenly over the surface.

5) When this application is dry, usually in about 15-20 minutes, re-charge the pad with your mixture and apply your second coat. Repeat this procedure as many times as necessary to make a build up of your mixture on the surface. Do not apply more than four coats in one day.

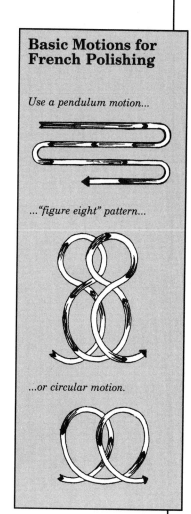

Basic Motions for French Polishing

Use a pendulum motion...

..."figure eight" pattern...

...or circular motion.

6) As additional coats are applied, add a drop or two of either linseed oil or olive oil to the charged rubber. This will act as a lubricant, making the movement of the pad over the surface a smooth operation rather than dragging and leaving marks.

7) The application of 8 to 12 coats is usually sufficient to produce a fine finish.

Ready made French polishing materials are offered today under such names as Pad Lac, Lacover, Qualasole, French Finish, etc.

French polishing is a skill that must be practiced, but once mastered it will produce a finish that will be the envy of those who see it.

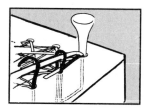

SECTION 4

Specialty Finishes-Repairs and Materials

The processes described in this section are not in the realm of this text. However, kits, materials and information for most of these can be readily obtained from Art and Craft Stores, Large Paint stores and suppliers of finishing materials.

STENCILING Stenciled designs are often found on old antique pieces of furniture, or you may wish to use them on your new creation. The technique of stenciling a design is a relatively simple procedure. Kits containing ready made pre-cut stencils are available complete with stencils, paints and brush, or you may elect to purchase the raw materials and cut your own unique stencils.

GILDING The application of Gold Leaf to a surface has been traced back to Biblical times and the art and craft is flourishing today; however, the tools and materials are more sophisticated and easier to use. Simply, the procedure is to prepare the surface, applying a sizing material, followed by the application of Gold Leaf and finished with an application of glaze and/or a protective coating. Since genuine Gold Leaf (23.5 carat) is expensive, but available, it is often substituted with an imitation Gold Leaf. Silver Leaf is also available.

CLEANING METAL HARDWARE During your restoration, you may elect to refinish the metal hardware that adorned your selected restoration. Metal polishes for various types of metal are easily accessible in the housewares department of your local grocery store. Or you can purchase metal restoration kits. If the hardware was plated and the plating has peeled off, it may be replated again or the entire piece replaced. Whatever the case, it will be to your advantage to protect your rejuvenated or new hardware with a clear protective coating.

CANING & RUSHING During the restoration of old furniture, it is often necessary to replace and install new cane or rush material. Cane itself is available in a variety of sizes for hand weaving or it can be obtained as pre-woven or machine woven which is easily applied to a seat having a groove, into which the cane is held in place with a spline. Rush material is also readily available in several diameters. This material must be woven by hand.

Booklets are available detailing the various designs and methods of using cane and rush.

CANING

RUSHING

FLOCKING You may encounter, with either new pieces or during restoration, the need for the application of a piece of felt (the bottoms of lamps), or Velvet (jewelry box linings), or Suede (drawer linings) or a similar fabric surface. To apply a piece of cloth to a surface can be very exasperating, messy and expensive; however the use of the "Flocking" technique provides an inexpensive, easy, satisfactory substitute. Simply, you apply a colored adhesive to the surface then blow the lustrous fibers onto this surface. When dry, you have a cloth covered surface.

DISTRESSING The processing of marking a surface to give it the appearance of having been in use for many years. This is easily accomplished mechanically by dumping a pailful of nuts, bolts, chain etc. onto the surface. Burning portions with a hot iron is another method. Speckling, using a stiff brush dipped into thick paint, then flecked with a comb is another method. The best method is to let your Grandchildren near it for about one hour.

DECOUPAGE Surfaces are sometimes decorated with photographs, drawings, maps etc. The process of incorporating them onto the surface is referred to as "Decoupage." The process is simple but you must follow specific technical details in a definite sequence.

PLASTIC DIP A heavy duty, opaque, water based flexible coating which seals out air and moisture and protects wood from rotting. Ideal for exterior wood surfaces such as lawn furniture and fences. Available for Dip, Spray or Brush application.

124

CARE OF
UNUSUAL SURFACES

REJUVENATING MARBLE SURFACES Slabs of Marble are often incorporated into a piece of furniture and these surfaces are also subjected to the same wear and tear as a wooden surface. The following tips can be used to rejuvenate a Marble surface.

● For stains, make a poultice of white blotting paper or white tissues and whiting (or ordinary household cleanser). Soak this poultice with the recommended solvent for the particular stain you are trying to remove.

   a.) Organic stains such as tea, coffee, cocoa, flowers etc. Use hydrogen peroxide or household ammonia for your poultice.

   b.) Oily type stains such as from meats, butter, hair oils, etc. Use lacquer thinner, acetone, or lighter fluid for your poultice.

● Place the poultice with the proper solvent over the stained spot. As the solvent evaporates, it will draw up the stain into the powder. Repeat if necessary.

1 *Minor scratches can be removed using very fine wet-or dry abrasive paper #600.*

2 *Rub with 4/0 Pumice.*

3 *Follow by rubbing with rottenstone. Use water as the lubricant.*

4 *When the surface is clean and smooth, apply a coat of paste wax or a good furniture polish.*

CARE OF A LEATHER SURFACE Damage to a leather surface requires the expertise of a qualified leather worker. Cuts, burns, abrasions and damage to a gold embossed area can be repaired, and information is available for making such repairs.

Use saddle soap or castile soap and prepare a solution to wash the leather surface. Rub the leather surface with cloth dampened, not soaked in the solution. Follow by wiping with a clean cloth and clear water. Dry thoroughly and apply a neutral leather dressing. Polish with a lambswool shoe brush.

Avoid the use of excessive water and strong soaps.

REJUVENATING CEDAR WOOD For those who own cedar closets or cedar chests and use them to keep the moth population away from their garments, the effectiveness of the natural aromatic cedar oil dissipates over the years and the worth of the cedar as a moth repellent is soon lost.

This characteristic of cedar can be revitalized by lightly sanding the exposed surface of the cedar wood and applying some cedar oil to this surface. A small amount of this oil applied to a clean cloth and rubbed into the surface is all that is necessary.

Keeping the chest lid or closet door closed helps to contain these active, aromatic vapors for many years.

GLOSSARY

ABRASIVES– A sharp hard substance, such as flint, garnet, silicon carbide and aluminum oxide, used to smooth wood surfaces.

ACRYLIC FINISH– A wood finish that can be thinned with water.

ADHESIVE– A natural or synthetic material that bonds two surfaces together.

ALLIGATORING– Numerous cracks in a film of finish. Caused by inflexibility of the finishing material, incompatibility between adjacent surfaces of different finishing materials, or improper surface preparation.

ANILENE DYE– Synthetic colorant used in wood stains. Formulated to dissolve in oil, water or alcohol.

BLEEDING– When the color of a coating material works its way up through succeeding coats.

BLUSHING– A white haze which forms, usually in a lacquer film, due to the film absorbing atmospheric moisture because of its rapid drying.

BOILED LINSEED OIL– Oil derived from Flax seed. It is heated, not boiled, and driers added.

BURN IN STICK– Also called shellac stick or lacquer stick. Available in a wide variety of colors. Used with a heated knife and, when molten, applied as a filler in cracks and small voids.

BUTTON SHELLAC– The least refined grade of shellac. Dark brown in color.

CLEAT– A strip of wood fastened to another piece to provide a holding or bracing effect.

COLORANT– Any material used to give color to something else. Often referred to as pigment.

COMPATIBLE– Able to bond with adjacent surface materials.

CRAZING– Thousands of tiny interconnecting cracks that can occur in a finish.

DANISH OIL– A penetrating oil finish made from a mixture of various oils, resins, solvents and driers.

DENATURED ALCOHOL– Ethyl alcohol with additives. Used as a solvent for shellac. Highly poisonous.

DYE– Any coloring agent; or the solution containing a coloring agent.

127

EPOXY– Synthetic resins used in paints and varnishes. Used with a catalyst. Produces an extremely hard and wear resistant surface.

EVAPORATION– To change into a vapor.

FILLER STICK– A wax-base putty available in a variety of colors, used to fill small small holes after a finish has been applied.

FISH EYES– Small round depressions in a finished surface cause by contamination with silicones.

FRENCH POLISHING– A method of applying multiple coats of shellac using a pad.

GEL STAINS & FINISHES– Stains and Finishes supplied in a Gel type vehicle which is easily applied to a surface with a rag.

GLUE– An adhesive substance used to join or bond different surfaces.

GLAZING– The application of a transparent or translucent coating applied over a painted or stained surface to produce blended effects.

JAPAN COLORS– Opaque colored pigments combined in a vehicle that is compatible with either oil or lacquer based products.

LACQUER– A tough, durable, brittle, fast drying finish.

LINSEED OIL– Oil obtained from the Flax seed. Raw linseed will not dry.

MINERAL SPIRITS– Petroleum based solvents often used as a thinner for paints, varnishes and polyurethanes.

NGR STAIN– Non-grain raising stain. Aniline dyes dissolved in methanol alcohol.

NITROCELLULOSE– The basic ingredient of most lacquers.

NON-BLOOMING RUBBING OIL– A water white, highly refined type of paraffin oil.

OIL STAIN– An oil base dye or pigment used to color wood.

OLEORESINOUS– Combination of oils and resins.

ORANGE PEEL– Pebbled appearance of a sprayed film due to its failure to completely flow level before drying.

ORANGE SHELLAC– A refined grade of shellac which still retains some of the orange-brown color of raw shellac.

OXIDIZE– Part of the drying process for some finishing materials.

PADDING LACQUER– A special type of lacquer applied in a thechnique similar to "French Polishing". Mostly used to repair damaged finishes.

PARAFFIN OIL– Mineral oil used as a lubricant when rubbing out a finish. Often referred to as Rubbing Oil.

PASTE WOOD FILLER– A mixture of silex (powdered quartz), boiled linseed oil, colorant, and driers. Used to fill pores of open grained woods.

PATINA– The condition of a wood surface and its finish that develops over time.

PENETRATING STAIN– A solution of solvent and dyes which penetrate into the wood surface.

PIGMENT– Insoluble finely ground powders mixed into a finishing material to provide body and color.

PIGMENTED STAIN– A solution of solvent and pigments which remain on the surface of the wood, rather than penetrate the surface.

POLISHING COMPOUND– A mixture of fine abrasive powder and a lubricant used for rubbing a surface.

POLYMERIZATION– Part of the drying process where the molecules of a substance interlock with each other.

POLYURETHANE– A synthetic "Varnish." Produces a very durable finish. Resistant to water, alcohol, weathering, wear and abrasions.

POULTICE– A soft moist mass.

PUMICE– A fine ground, white, abrasive powder used for rubbing a finish.

ROTTENSTONE– A fine ground, dark, abrasive powder, used for rubbing a finish. Finer than pumice.

SCHEDULE– A step-by-step chronological procedure using a variety of finishing materials.

SEALER– A finishing material used to seal the pores of bare wood. Also used as a coat between two incompatible finishing materials.

SCHEDULE– A step-by-step procedural guide.

SHEEN– The degree of light reflectiveness from a finished surface.

SHELF LIFE– The maximum time a product can be stored in a closed container prior to initial use.

SHELLAC– A thin solution of shellac flakes dissolved in denatured alcohol.

SIZING– A material applied to a surface to fill the pores and reduce the absorption of subsequent materials into the wood.

SOLVENT– A liquid which will dissolve other materials.

SPAR VARNISH– A tough, transparent, waterproof form of varnish. Recommended for exterior use.

STAIN– Any material that artificially colors wood without obscuring the grain or texture.

STRIPPER– A combination of chemicals and solvents used to soften and loosen an old finish for removal.

TACK CLOTH– A piece of cheesecloth that has been treated to attract dust.

THINNER– A liquid used to control the consistency of a finishing material.

TOXIC– Describes a material that may be harmful, destructive or deadly.

TRICHLOROETHANE– Thinner used for Veneer Glue.

TUNG OIL– A natural oil from the Tung Tree. Used by itself or in combination with other oils to make a penetrating type stain or finish.

TURPENTINE– Distilled gum from pine trees. Used as a solvent for oil based finishing materials.

UNIVERSAL COLORS– Tinting colors that are compatible with oil or water based products. Not compatible with lacquer, epoxies or catalyzed finishes.

VARNISH– A clear transparent mixture of either natural gums or resins and oils, which dries by oxidation.

VEHICLE The liquid part of a finishing material.

VENEER– A thin piece of wood.

WASH COAT– A thin coat of finishing material that seals or locks in other finishing materials or wood pores.

WATER STAIN– Aniline dye that is dissolved in warm water. Its use will raise the grain of the wood.

WHITE SHELLAC– Most highly refined grade of shellac. Bleached to remove all color cast of the raw shellac.

ACKNOWLEDGEMENTS

The author is deeply indebted to many people for making this book a reality, especially:

The many students who over the years encouraged me to put my words in writing.

The executive staff of Constantines, Mrs. Gertrude Constantine, Mrs. Dorothy Docherty and Mr. Glenn Docherty. Their confidence, encouragement and assistance can not be surpassed.

Jane Corcillo and Elaine Cheh for preparing the written text, photographs, charts and drawings into a form suitable for publication.

Seth Goltzer for his superior efforts with the camera.

The friends and relatives who edited the original manuscript.

And finally the many individuals and industrial firms that generously supplied me with photographs, drawings, charts and the latest available technical information pertaining to their products. This list includes the following:

Adjustable Clamp Co. (Clamps)

H. Behlen & Bro., Inc. (Finishing Materials)

Borden Inc. (Wood Filler, Elmer's Glue)

Central Rubber Products Co., Inc. (Glue Applicators)

Charles B. Chrystal Co., Inc. (Pumice, Rottenstone)

Clesco Mfg. Divn. (Drum Sanders, Drill stop collars, Dowel Centers and Mandrels)

Albert Constantine & Sons, Inc. (Woodworking Supplies)

Deft, Inc. (Finishing Materials)

Diamond Machining Technology Inc. (Diamond Stones)

Donjer Products Co. (Suede Tex)

EMCO Electric Co. (Electric Glue Pots)

Forest Products Lab. (Wood Finishing Information)

Franklin International (Glues)

General Finishes Sales & Service Corp. (The Sealacell System)

General Hardware Mfg. Co., Inc.

The Otto Gerdau Co. (Cane etc.)

Gill Chemical Products (Mend-All)

The Hope Co., Inc. (Finishing Materials)

Walter Kidde (Fire Extinguishers)

Klean-Strip, Divn. (P & V removers)

M.W.B. Industries Inc. (Tack Cloths)

Marshall Imports (Antiquax)

Minwax Co., Inc. (Finishing Materials)

E. C. Mitchell Co., Inc. (Abrasive Cords & Tapes)

Norton Co. (Abrasives)

PDI, Inc. (Plastic Dip)

Portalign Tool Divn. (Drill Guide)

Porter-Cable Corp. (Hand Power tools)

Pratt & Lambert - M. L. Campbell (Finishes, Fabulon)

Roger A. Reed, Inc.

Sansher Corp (Stripping Material)

Shoplyne (Safety Products)

Stanley Tools

Star Chemical Co., Inc.

The Slomons Group (Adhesive "Woodwiz")

WATCO Dennis Corp.

Wm. Zinsser (Shellac)

BIBLIOGRAPHY and REFERENCES

Miller, Robert S.-Home Construction Projects with Adhesives and Glues

Hobbs, Harry J.-Veneering Simplified

Hobbs, Harry J. /Fitchett, Allan E.-Modern Marquetry Handbook

Constantine, Albert- Know Your Woods

Soderberg, George A.-Finishing Technology

Kinney, Ralph-Complete Book of Furniture Repair & Refinishing

Newell/Holtrop-Coloring, Finishing and Painting Wood

Stieri, Emanuele-Woodworking for the Home Craftsman

INDEX